LIVING IN HERITAGE

MATERIAL VERNACULARS

Jason Baird Jackson, editor

LIVING IN HERITAGE

Tulou as Vernacular Architecture, Global Asset, and Tourist Destination in Contemporary China

—⠍—

LIJUN ZHANG

INDIANA UNIVERSITY PRESS

This book is a publication of

Indiana University Press
Office of Scholarly Publishing
Herman B Wells Library 350
1320 East 10th Street
Bloomington, Indiana 47405 USA

iupress.org

Manufactured in the United States of America

First Printing 2024

Indiana University Press is pleased to make this monograph freely available as an Open Access monograph. To read or download, visit https://hdl.handle.net/2022/29905.

Cataloging information is available from the Library of Congress.

ISBN 978-0-253-07096-8 (hardback)
ISBN 978-0-253-07097-5 (paperback)
ISBN 978-0-253-07098-2 (web PDF)

CONTENTS

Acknowledgments *vii*

Note on Romanization, Translation, and Names *ix*

Introduction *1*

1. Localness in the Time of Great Transformation *19*

2. Tulou as Vernacular Architecture *47*

3. Tulou as Home and Lived Experience *74*

4. World Heritage Nomination and the Institutionalization of Tulou *93*

5. Everyday Engagement with Heritage Process and Tourism Activities *116*

6. Tulou as New Resource and Power Leverage *137*

Conclusion *152*

Bibliography *159*

Index *169*

ACKNOWLEDGMENTS

THIS BOOK WOULD NOT HAVE BEEN possible without the support and help of many people, all of whose names I could not mention here. I am using this opportunity to express my gratitude to them.

I would like to express my heartfelt gratitude to those who were engaged in my ethnographic research in various forms, especially the community members in Hongkeng Village who allowed me to enter their lives and living spaces. Expressions of thanks must begin with Lin Qiyi's and Lin Maoxin's families in Yucheng Building in Hongkeng Village. They hosted me, helped me settle in the village, connected me to various community members, and allowed me to observe and participate in both their daily activities and special occasions, such as festivals and wedding ceremonies. I treasure the time I spent and relationships I formed with the residents in Yucheng Building. We ate together every day, sat and chatted on the long bench in the hallway or under the eaves after lunch and dinner, visited community members and relatives, drank tea with the tourists, and walked around the village. The experience was essential for me to learn and understand what is presented and discussed in this book. In 2023, I got the chance to revisit Hongkeng Village many years after I finished my last field research in the community. It was heartwarming to hear my host family members also share memories of the time we spent together.

Numerous other people in Yongding County helped me in various ways in the process of the project. Thanks go to all of them but especially to the many consultants I met and talked with during my fieldwork, including Lin Hengsheng, Lin Dengrui, Lin Jianchan, Jiang Jinrong, Lin Rigen, Lin Fuyuan, Liu Zenjing, Su Ziqiang, and Zhen Xincai. I spent long hours talking with them about tulou, the village history and culture, their personal experience, the process of heritage nomination, and tourism development. I would like to thank my cohort Liu Xiaochun at Beijing Normal University and her family in the Yongding County seat. They welcomed me and hosted me during my fieldwork in the county seat.

This book is a revised version of my doctoral dissertation completed for the Department of Folklore and Ethnomusicology at Indiana University. My deep appreciation also goes to my advisers and mentors at Indiana University. I am thankful for their aspiring guidance and invaluably constructive criticism and advice at various stages of this project. Among them, I owe Jason Baird Jackson and Michael Dylan Foster the deepest gratitude. They scrupulously and critically read every page of the original version of the manuscript and offered critically constructive comments and suggestions. I also thank Pravina Shukla and Sara Friedman for their support and feedback. Jason Jackson is the editor of the Material Vernaculars book series. He provided tremendous help and advice for the revision and preparation of the book manuscript. I give particular thanks to him for his thoughtful leadership and direction.

I am grateful as well to my colleagues in the Folklore Program at George Mason University—Ben Gatling, Lisa Gilman, and Debra Lattanzi Shutika—for their help and support.

I thank the two anonymous reviewers for their careful and insightful comments. The suggestions they offered were very helpful for me in improving the final manuscript.

Thanks to Gary Dunham and Indiana University Press for the wonderful support of the book project. And thanks to Allison Chaplin, Anna Garnai, Sophia Hebert, Darja Malcolm-Clarke, Vinodhini Kumarasamy, Samantha Heffner, and many other people in the publication team of Indiana University Press for their dedicated work on the publication of the book.

NOTE ON ROMANIZATION, TRANSLATION, AND NAMES

THE PRIMARY LANGUAGES of fieldwork are Standard Chinese (Putonghua, or Mandarin), which is the national language of the People's Republic of China, and local Hakka dialect (Kejiahua), which is the dialect spoken by the Hakka people living in my field sites. Most of the interviews and conversations were conducted in Mandarin Chinese. Some interviews were conducted in Hakka dialect, especially in cases where consultants were senior in age and not able to speak fluent Mandarin Chinese. The transcriptions are in Mandarin Chinese.

Chinese terms are romanized according to the Hanyu pinyin system and are denoted by italics at their first occurrence in the book. There are four tones for Mandarin. I have chosen not to indicate tones in the text. The Chinese terms are presented in both pinyin and Chinese characters.

To protect the consultants' privacy, I chose to use pseudonyms for community members and the officials whom I interviewed. In narrative passages, individuals with Chinese language personal and family names are consistently named in the normal Chinese order, with family name given first.

LIVING IN HERITAGE

—⚉—

INTRODUCTION

Almost three years after UNESCO (the United Nations Educational, Scientific, and Cultural Organization) designated the massive communal dwelling type known as *tulou* (earthen building or rammed-earth building, 土楼) as World Heritage, on a drizzling morning in late spring of 2011, I boarded a bus from Longyan, the prefecture-level city and political, economic, and transportation center of the region, to Hongkeng Village, a once remote rural community in western Fujian Province in Southeast China. Before boarding, I looked where the destination names are marked on the windshield to verify that it was the correct bus. I was immediately struck by seeing the two bright yellow Chinese characters of "tulou" indicating our destination. Unlike the other buses, whose destinations were marked with the names of the city and village locations, my destination was marked by the name of a form of vernacular architecture. For this first visit to Hongkeng Village, my sister offered to go with me, partly because she wanted to keep me company, partly because she wanted to take this opportunity to visit a tulou "scenic spot" (Nyíri 2006). Even though she had seen several tulou as residential houses in the town where she worked, she hadn't visited any of the tulou heritage sites that had become popular tourist destinations.

Through slopes of dense trees and bushes, the bus traveled along the two-lane road with frequent ups and downs and abrupt twists and turns. Towns, villages, and single houses were scattered throughout the valleys, with patches of tobacco and rice fields lying at the bases of hills and mountains. Along the way, numerous signs reading, "Tulou Scenic Spots," with indications of the direction

Figure 0.1   Tulou Visitor Center. (Photo by author)

and distance remaining, led us toward our destination. Along the way outside the bus window, slogans, promotions, and advertisements printed on billboards passed by. Many of them had pictures of tulou as their backgrounds. Among them, the most conspicuous sign was a billboard approximately ten feet high and three feet wide in the center of a township about eighteen miles away from our final destination. It presented a picture of former president of China Hu Jingtao visiting tulou and was inscribed with the words he publicly delivered during the visit: "Hakka tulou is a treasure of Chinese culture. It exemplifies the harmonious relationship of a large family living under the same roof. [I] hope the residents will preserve, inherit, and make full use of the heritage."

After we had been bombarded with these tulou signs and images for about two hours, the bus stopped in front of a massive building with beige walls and a black roof, a new construction that was designed to look like traditional tulou. I later learned that this was the Tulou Visitor Center (fig. 0.1). The bus driver, knowing the passengers' purpose of the visit, pointed to a building located on the other side of a big parking lot and told us it was the place to buy tickets for

Figure 0.2 Entrance to Hongkeng Village in Hukeng Township of Yongding County, Fujian Province. (Photo by author)

admission to the Hongkeng tulou scenic spot. Instead of following the other passengers toward the ticket center, I called Lin Guya, a contact I obtained through a personal connection. Guya was a leader of the Hongkeng Villager Committee and also worked for the local tourism management company Hongkeng Hakka Tulou Tourism Corporation Ltd.[1] He didn't expect me to get off the bus at the recently constructed Tourist Service Center since the actual entrance to the village was still more than half a mile away. Nevertheless, he came to meet me at the visitor center and introduced me to his village.

Behind the actual entrance to the village—marked by a grand decorated archway with shining golden characters reading "Yongding Tulou Folk Culture Village" on the beam—a barrier blocked the road (fig. 0.2).[2] A security man in a blue uniform stood in front of a beige security office next to the barrier. Only visitors with tickets or vehicles with special permits were allowed to enter the village. On the street in front of the archway was a small business corner. Restaurants and hotels competed for customers by having local people wearing bamboo hats deliver business cards to visitors.

Figure 0.3 Exterior view of Yucheng Building in Hongkeng Village with tourists experiencing picking up persimmons with a homemade tool in the front yard. (Photo by author)

In the car taking us into the village, Guya bluntly asked me the purpose of my visit. I told him I came to do my field research and hopefully to find a place to stay in the village. He kept silent after this very brief exchange. From what happened after we got off the car, I later realized that in this short period of silence he had effectively processed the information I provided to him and knew where to take me. About a minute after I had communicated my intentions, the car stopped. As I got out, I found myself standing in a yard paved with cobblestones. In front of me was a rectangular tulou with black characters painted above the thick wooden gate that marked it as the Yucheng Building (fig. 0.3). On each side of the gate hung large round bamboo trays with the characters informing people that this building is a hotel and a restaurant. Hanging on the wall on the left side, four smaller baskets with characters in red marked the building as a workshop for rice wine. Piles of wine jars were lined up against the beige wall under the signs. Tourists were posing and taking pictures in front of the jars. The baskets and wine jars are daily utensils creatively used by the house owners for decoration and as commercial signs to attract tourists' attention. I found such decorations and signs representing the traditional lifestyle of an agricultural society everywhere when I walked around the village later on. It had become a new tradition practiced by the villagers while the village was transitioning from an agriculture-based rural society to a service-oriented tourist site due to the development of heritage tourism.

Yucheng Building is a medium-sized tulou with three stories in the front section and four stories in the back section. According to the wooden plaque hung on the wall of the hallway with an introduction, in Chinese and English, of the building, it was constructed in 1875 and covers an area of about 1,500 square meters (16,416 square feet). As I followed Guya into the building, I passed a hallway and turned to the left at a passage surrounding the courtyard at the center of the building. A door connecting the hallway and the rooms on the left side of the passage marked the section that was separated from the right side of the building. Just past a small kitchen along the passage was an open-space living area facing the courtyard. A man in a white shirt greeted me with a welcoming business-style smile. He was introduced to me as Ziwu, one of the hosts of Yucheng Building who owns the left side of the building with his parents, his two brothers, and their families. I later learned that Ziwu, like Guya, was also one of the village cadres. In addition to their personal friendship, Ziwu's family is considered to be relatives of Guya's family, according to the village lineage relationship. Ziwu's extended family runs the daily business of selling Hakka rice wine in the building. But only his parents, his younger brother, Jianwu, and Jianwu's immediate family were permanent residents of the building. Ziwu and his own nuclear family and his older brother, Congwu, and Congwu's nuclear family had moved to the relocation houses recently constructed by the government outside Hongkeng Village. Ziwu and Congwu as well as their nuclear families came to the old tulou of their extended family every morning for daily business activities and went back to their new homes at the end of the day.

Ziwu ushered us into the living room for tea. We sat around a long wooden tea table, and as Ziwu carefully poured tea into tiny white porcelain cups, Guya introduced me and informed him of the purpose for my visit. Immediately after that, Ziwu yelled out a name in local Hakka dialect toward the opposite side of the building across the center courtyard. A young man in his early thirties came to us through a crowd of tourists, holding a menu in one hand and a pen in the other. I was introduced to the man named Binghan, the host of the other half of the building and also the owner of Yucheng Hotel and Restaurant.

After acknowledging that I would like to find a place to stay in the village, Binghan simply told me to go by myself and check the guest rooms on the fourth floor at the back of the building. Through the narrow wooden stairs next to Binghan's living room, I climbed to the top floor of the four-story building. There were two single-bed and two double-bed guest rooms. I was told there used to be five rooms on this floor that belonged to Binghan's family. A room in the middle was remodeled and transformed into three bathrooms. Binghan and his wife, Lizhen, told me that they would like to keep the rooms with attached bathrooms for their future customers. Bedrooms with attached modern bathrooms were rare in the village; thus, they were particularly popular among the tourists who came to experience the countryside living style but didn't want to give up the convenience provided by modern living facilities. The bathroom, as advertised on the board in the hallway of the building, became a selling point for their hotel.

In 2008, when tulou were designated as UNESCO World Heritage Sites, Binghan and Lizhen decided to come back to their inherited home—Yucheng Building—to start this heritage-inspired tourism business after having spent many years in the nearest metropolitan city, Xiamen, as city workers. Binghan adopted the building name for his restaurant and hotel. Binghan's parents lived with them in the building, helping them run the business and care for their two little children. The family of Binghan's elder brother, Dehan, like Ziwu's and his brother's families, had moved out of Yucheng Building and lived in the new relocation community. Dehan came back to his old home almost every day to help his brother with his business. And his wife, Suyun, worked as a local tour guide; when she was not showing tourists around the village, she spent her time in Yucheng Building.

I later became very close with my hosts Binghan and Lizhen and the other residents in the Yucheng Building. They allowed me to almost entirely immerse myself in their daily lives. However, during our first encounter, I found the reception was a mixture of habitual hospitality and commercial rationale in the building, which is a merger of living space and commercial space. What I saw and experienced on my first day in Hongkeng Village is a snapshot of a very small but vivid part of the heritage phenomena in the village specifically and in other parts of rural China more broadly.

As demonstrated, tulou are signs of value in the locale, and as a cultural form, the architecture style gives a specific identity to the region. The promotion of tulou is practiced by the local governments, the tulou residents, and commercial agents for various purposes. Even before arriving at the actual tulou "scenic spot," the adventure tale already had begun with the passengers' expectations of the site as well as the images and signs of tulou and World Heritage Sites along the way. Upon my arrival at the village, I was brought into a space characterized by highly self-conscious cultural presentations and expressions in staged scenes or curated spaces, such as the grand gateway, the creative basket signs, and the decorative wine jars.

ABOUT THIS BOOK

The study detailed in this book is based on ethnographic fieldwork conducted in three stages—2011, 2012, and 2014—with most of these periods spent in Yongding Hakka Earth Building Folk Cultural Village (a.k.a. Hongkeng Village) but also conducting short-term fieldwork in neighboring tulou heritage sites and the county administrative center, Fengcheng Town. My selected research site, Hongkeng Hakka Earth Building Folk Cultural Village, used to be primarily an agricultural community dependent on growing rice for subsistence and a few cash crops, such as tobacco, tea, and persimmon. After being designated a UNESCO World Heritage Site, this village of around 2,800 residents has turned into what tourists call a "living museum." This once relatively isolated community is now much more fully exposed to, and connected with, the outside world. With a large number of Chinese and international tourists coming, as well as the current and continuing reframing and construction of the residential village as a folk culture village and World Heritage Site, this village is transforming from an agriculture-based community to a service-oriented destination/community. Hongkeng Village is an ideal field site for a close examination of people's daily practices in an increasingly tourism-saturated and heritage-minded setting.

I also conducted shorter-term fieldwork in Fengcheng Town, where I collected another body of field data about the structure of the county governance, the government's motivation of tulou heritage nomination and tourism development, and the county's ambitions of, and planning for, modernization. Fengcheng Town, a small city about a one-hour drive away from Hongkeng Village, is where the county government and many other governmental agencies and institutes are located, including the headquarters of Yongding Hakka Tulou Tourism Company Ltd. and Yongding Bureau of Cultural Relics, which was formed on the basis of the Office of Yongding World Heritage Nomination Committee and is in charge of cultural heritage documentation and conservation in the area.

Intermittently over the course of my field research period, I conducted library and archival research in local repositories. As a tradition in Chinese community documentation, local gazetteers (*difang zhi*, 地方志) chronicling communities have been regularly written and updated by area scholars and elites and are currently well preserved in archives all over China. This genre of information and documentation is a rich resource for the understanding of local culture. Other historical documents that I collected and explored include lineage genealogies, stone inscriptions, liturgical and scriptural texts, village-based ritual and dramatic performance texts, and official documents. These materials provided me with provisional knowledge of community history, cultural activities, changes in population, economic statistics, and current development policies. The analysis of such documents helped me gain a fuller understanding of the local community's past and how it relates to the current sociocultural situation in the context of global heritage tourism and national cultural/economic policy.

My study of vernacular culture and heritage process as linked to the UNESCO recognition and to China's rapid socioeconomic transformation is centered on the Hongkeng villagers' physical homes, structures known as tulou. This project ethnographically documents the vernacular architecture and everyday life of Hongkeng Village as a rural community in Southeast China while it is engaging in (and with) heritage policies, heritage discourse and practices, and heritage-related tourism activities. The tulou architectural form has been through a process of transformation and transvaluation since it was turned from residential dwellings to global asset and further became a tourist destination. Focusing on tulou as a national phenomenon earning UNESCO-endorsed World Heritage designation and the recognition of Hongkeng Village in Fujian Province as a high-profile heritage space, this ethnography explores how the concept, discourse, and practice of heritage is localized in the traditionally agriculture-oriented and lineage-based village and how the tourism development brought by the heritage designation impacts tulou and community life. This ethnography also investigates how people in the community experience and negotiate tradition, identity, everyday economics, cultural representation, social relations, and property ownership in the wake of great cultural change and socioeconomic transformation that accompanies heritage-making, the implementation of heritage policies, and the development of cultural tourism.

This book examines both top-down and bottom-up heritage processes. Heritage is woven into the vibrant living traditions and creative expressions in this local community. Heritage discourse and practice intersects and interacts with local everyday life and the living world. I present and analyze how the heritage process is initiated and led by the institutional and authorized agencies, highlight the impacts of the heritage regime on local tulou residents' everyday experience, and examine the local people's agency and their active engagement with the heritage process. On one hand, the designation of tulou as World Heritage and the practices following the designation fundamentally contribute to shaping and transforming Hongkeng from a living space with low metacultural awareness to a highly regulated heritage and exhibition space with a high level of metacultural awareness (Jackson, Müske, and Zhang 2020). On the other hand, Hongkeng villagers' traditional culture and daily practice constitutes and constructs the local conception and practice of heritage. The heritagization and self-representation practiced by communities and individuals in Hongkeng Village and the neighboring municipalities is a process of bottom-up heritage-making on the ground that coexists, intertwines, and interacts with, as well as sometimes contests, the top-down heritage policies, discourses, and practices dictated by the state.

Local circumstances interact with the heritage regime in dynamic and complex ways. When Hongkeng residents encounter heritage policies and practices, and when their remote and relatively isolated agricultural village is resituated in the global system of world heritage, the local community (as well as its culture, history, and daily life) increasingly entangle with governmental management, market forces, and daily politics revolving around heritage. This entanglement engages and prompts continuous presentation, representation, and negotiation in local people's daily lives. In the realms and intersections of cultural expression, identity recognition, and daily economy, this entanglement also accelerates the changes and production of signs, meanings, and values.[3]

As part of the heritage process in China, the case of Hongkeng Village needs to be situated and understood in the context of national heritage policies and practices. In the following, I provide an overview of heritage policies, discourse, and practices as well as social campaigns and processes such as "new countryside construction" (xin nongcun jianshe, 新农村建设) in China to contextualize this on-the-ground study of the internal and external processes of the heritagization of the vernacular architecture tulou and a rural ethnic community undergoing social, economic, and political change at the hands of national and global forces.

HERITAGE AND SOCIAL DEVELOPMENT IN CHINA

The UNESCO Convention for the Safeguarding of the Intangible Cultural Heritage (ICH) was adopted at UNESCO's thirty-second session in Paris in 2003. In the following year, China signed the convention and became the sixth nation to join the international endeavor of safeguarding ICH, which is regarded by UNESCO "as a mainspring of cultural diversity and a guarantee of sustainable development" (UNESCO, n.d.-d). Ever since, the convention, including the explanation and measures of safeguarding, has served as the most important framework and instrument for ICH discourse, practice, and policymaking in China. Scholars, policymakers, and cultural practitioners refer to the convention as an instrumental tool to guide and legitimate the practice of ICH identification, research, safeguarding, promotion, and other activities, with a large and growing body of scholarly work on translating, interpreting, analyzing, and reflecting on the texts, policies, and spirit of the 2003 convention (Chao 2016; Bamo 2008, 2015; Bamo and Zhang 2016; Gao 2017; Ma 2017; Liu 2009; Zhang 2019).

Ever since China joined the 2003 convention, people have witnessed the large scale of ICH practice and its dominance in the scholarly and social discourse in China.[4] Various actors, including scholars, cultural practitioners, and political advocates, tour agencies, and business people are actively involved in heritage-related discourse and practice from the national level to the local level across the nation. This phenomenon is set against the broader background of dramatic social transformations that have taken place since the start of China's "reform and open up" period (1978–present). China has experienced rapid economic, technological, and infrastructural development along with radical social transformations, which greatly impact people's daily lives and traditional practices. In the urban areas and more prosperous rural areas, people have widely adopted modern lifestyles, and many customary practices have been or are being simplified or changed. In this process of urbanization and industrialization, nostalgia and concern about the current condition and the future continuity of traditional culture becomes a powerful force to promote heritage designations and cultural conservation (Liu 2016; Yang 2020; Zhu 2017).

Heritage is often being transformed into cultural, economic or political capital. In the context of heritage tourism, local culture has become an important resource for economic development and a source of political mobilization. It has become a common practice in China, and around the world, to transform a heritage site into a destination that attracts visitors to consume the heritage as well as associated local goods and services. Heritage is a way of producing a destination. As Barbara Kirshenblatt-Gimblett (1995, 371) points out, "Tourism and heritage are collaborative industries, heritage converting locations into destinations and tourism making them economically viable as exhibits of themselves. Locations become museums of themselves within a tourism economy." The practices of tourism transvalue and transform heritage in the process of performance, exhibition, representation, advertising, and touristic management. Heritage becomes a force that mobilizes people and resources, reforms discourses, and transforms practices (Hafstein 2007, 75).

In the economically underdeveloped but culturally more diverse areas, traditional or ethnic culture is oftentimes promoted as an important resource for economic development and a strategy for sustaining the vitality of rural community within the state's campaign to eliminate poverty and promote "new countryside construction."[5] In this context, heritage is employed as cultural and economic capital for both the tourism industry and the "cultural industry" (*wenhua chanye*, 文化产业) or cultural economy at various scales and by different agencies within China while the nation pursues modernization and addresses issues such as the imbalance in economic development between different populations and regions (Chan 2005; Fan 2016; Nyíri 2006; Oakes 1998; Tian 2014; Wang 2020).

Internationally, China has emerged as an important member of the global interstate system and has been more active on the global stage. Concomitant with China's more frequent encountering and interaction with the world, we witness a new wave of Chinese nationalism and a more enthusiastic pursuit of international recognition and legitimation. In this context, ICH is perceived as an important source of national pride. Moreover, joining the UNESCO convention and having "items" of traditional culture and folklore designated as ICH in the UNESCO lists can serve as a path to international cultural exchange and communication, bridging intercultural understanding, presentation of national culture, promotion of cultural diversity, and creating opportunity for international collaboration (Zhu 2017).

All of these factors contribute to the prevalence of the heritage-related discourse and activities in a national context that creates new meaning and conception around ideas of urban/rural, east/west, north/south, and rapidly developed/underdeveloped as China becomes more active on the global stage. The heritage phenomenon in China is so prevalent that it is characterized as "heritage fever" or, less dramatically, "heritage movement" by scholars (Peng 2008; Wang 2008; Zhang and Zhou 2017). As folklorists Juwen Zhang and Xing Zhou (2017, 139) noted, "the ICH movement in China has not only provided a 'timely' venue for alleviating domestic problems, but has also given an international platform on which China is able to stand with its unique ICH concept and political and academic discourses." The 2003 convention, in addition to the UNESCO Convention Concerning the Protection of the World Cultural and Natural Heritage adopted by UNESCO in 1972, plays a convenient role in legitimating and authorizing the national heritage discourses and practices while China faces the issues of development, urbanization, migration, rural and urban economic imbalance, cultural change, connection to the world nationally and in the larger global context, and its interstate system.

ICH SAFEGUARDING AND SOCIAL DEVELOPMENT AS GRANT NARRATIVE AND WORKING GUIDELINE

According to the official guidelines for ICH safeguarding issued by the General Office of the State Council, the major goal of ICH safeguarding is to establish an ICH safeguarding system with "Chinese characteristics" through the effort by the whole society to "effectively" protect the "precious, endangered ICH that are of historical, cultural, and scientific values" and to ensure the transmission and development of these heritage areas. The government perceives "effective protection and use of ICH" as important for achieving the "holistic, balanced, and sustainable development of economy and society" (General Office of the State Council 2005). The guideline is an official document that is repeated, adopted, interpreted, and reinterpreted in the ICH policies and practices from the national level to the local level within the institutional system of ICH safeguarding. The connection between ICH safeguarding and economic and social development is a grand narrative that dominates and facilitates the on-the-ground heritage discourse and practice. The narrative is often incorporated into or put in juxtaposition with other grant narratives and social development blueprints, including the current social campaigns of "poverty alleviation" (*fupin*, 扶贫), "construction of new countryside," and "innovative urbanization" (*xinxing chengzhen hua*, 新型城镇化) as well as the "Belt and Road Initiative" (*yidai yilu*, 一带一路) (Guo and Li 2019; Han and Gao 2020; Xie 2014; Zhu 2017).

The working guideline or policy for ICH safeguarding is stated in the document as "giving priority to safeguarding, putting rescue first, using (heritage) in an appropriate way, (ensuring) transmission and development." This becomes a slogan that is commonly seen on banners across the nation and in various other media. Echoing the goal of ICH safeguarding stated previously, the guideline emphasizes the significance of ICH safeguarding, or in some more urgent cases, ICH rescue, and at the same time, authorizes the use of ICH for broader purposes. In heritage practice, this is often used as evidence of institutional legitimacy or justification for employing ICH as a resource in the economic and commercial sphere. The document further explains that people should "put the relationship of the safeguarding and use of ICH in the right place," using ICH in the condition that "it is effectively safeguarded in authentic and holistic ways." It also warns against ICH being "misused or overused." The working guideline emphasizes that, based on the "scientific identification" of ICH, institutions should apply effective measures to ensure that ICH is "acknowledged, respected, and promoted by the whole society" (General Office of the State Council 2005). Although the actual interpretation and implementation of the guideline is subject to the choices and activities of policymakers and cultural practitioners, the guideline reflects the implication of safeguarding in the sense of conservation, indicating the involvement of management that controls practitioners' behavior while allowing the use of what is designated to be conserved. In theory, to ensure sustainability, the approach requires the subjective will to use resources wisely in a controlled manner. But in reality, it might not always be the case, and the approach may risk cultural sustainability and diversity. Recent ethnographies of heritage activity in China (Chio 2014; Luo 2020; Nitzky 2013; You 2020; Zhang 2020) and elsewhere (Foster and Gilman 2015) have sought to track the unintended consequences arising from this gap between policy and practice.

The heritage practice in China in the institutional sense is essentially government-led (Tian 2009; Zhou 2012; Tang 2014; An and Yang 2015). Heritage practice on the national level is a process of institutionalization that involves national efforts in joining international heritage-related organizations and conventions while developing national institutes and administrative agencies for heritage-centered activities (Zhang 2019). In the Chinese institutional system, the Ministry of Culture and Tourism, which is most directly connected to UNESCO on the national level, functions as the leading authorized institution for heritage nomination and safeguarding policy and practice. Heritage and "tradition carriers" are granted special designation by four levels of government: the central government, the provincial government, the municipal government, and the county government (Chen 2010; Kong and Song 2018).[6] Although community participation has been emphasized in the ICH-related documents by both UNESCO and China, in actual practice, safeguarding decisions and related policymaking are usually administrated by these official agencies. The actual implementation of the policies and practice of safeguarding framed by the UNESCO conventions may result in an imbalanced subject-object power relationship between the conservators (whether literally, as technical specialists like museum conservators, or figuratively, as with political leaders who formulate conservation-focused policy) and the local social groups in which heritage forms and practices exist (An 2016; Tian and Liang 2020; You 2020). In the following chapter, I closely examine these dynamics, looking at the local historical context and social development over time. I explore the knowledge and practices surrounding the materiality of the tulou architectural style and tulou residents' communal dwelling practices, the transformation of tulou from residential homes to UNESCO World Heritage Sites, and local people's everyday engagement with the heritage process centered on tulou. Finally, I discuss how tulou have become a local resource and source of leverage as World Heritage Sites as well as how heritage tourism reconfigures broader social forces in China.

GUIDE TO CHAPTERS

Chapter 1 puts tulou and Hongkeng Village in the network of local natural environment, history, social circumstance, cultural tradition, and economic development. The contextualization helps us understand that tulou and related heritage and tourism are conditioned by local time, space, knowledge, and social relationship. Tulou, as a particular vernacular architecture style that is only to be found in a confined area in Southeast China, is an integrated part of the local community. The local geographical features, historical development, and social-cultural situation are significant formative factors from which tulou have emerged and developed as a form of vernacular architecture. Meanwhile, the distinctive architectural form of tulou and local community life that formed through historical processes supports the World Heritage nomination and, furthermore, becomes a significant resource for the presentation of localness in heritage tourism. At the same time, the modern use of tulou as political, cultural, and economic resources is situated by the local government's ambition, ideology, and related political practices discussed in this chapter.

Connecting to the natural, historical, and sociocultural contexts describe in chapter 1, chapter 2 focuses on the tangible and intangible aspects of tulou as a form of vernacular architecture. I examine not only "the five properties" of tulou—namely, history, material, construction, design, and function—but also its relation to local culture and society (Fleming 1974, 153). The local natural resources, skills, traditional knowledge, customs and beliefs, and social system have shaping influence on the design, construction, and use of tulou as large, multistory, multifamily vernacular buildings enclosed by thick rammed-earth walls. This chapter also illustrates how community members negotiate with their living environment, space, and social relationships in the process of designing and constructing tulou. The tangible, material or physical features of tulou reflect and convey its residents' daily lives as well as the characteristics of the local culture, social relationships, and belief system.

Chapter 3 looks at the continuity and change of tulou residents' customary daily activities, social lives, and cultural practices. Tulou residents form a sense of place and social space through not only personal life and experience but also shared experiences and collective memories of many generations. Despite growing ambivalence about tulou as contemporary residences, tulou not only meet local people's residential needs, as their private homes and as the space where people experience "the rites of passage," but also function as the physical signs that represent social structure and relationships. Hongkeng Village is a lineage-based society. Tulou is a physical demonstration of the lineage system and its development and thus is important to local people for understanding their roots and their present position in the network of social space.

Chapter 4 analyzes the transformation of tulou from residential home hidden in the valleys of Southeast China to a recognized site of world heritage and global asset. Tulou heritage nomination is part of the ongoing national and global heritage process and discourse, but it is also a localized practice. In this local endeavor to enter the global cultural and institutional system, authoritative forces such as local government and cultural experts play major roles while community members' agency is limited. Multinational bodies, governments, organizations, scholars, and corporations mediate and heavily influence local understanding and conception of heritage. In addition, as demonstrated in chapter 4, the heritage nomination work requires enormous investments of time, personnel, and funds, which are all expected to produce a return in the form of future profits, usually through the development of heritage tourism.

Chapter 5 investigates community members' everyday engagement with the heritage process and tourism activities as well as the museumification of domestic and public space in Hongkeng Village. The designation of tulou as World Heritage and the development of heritage tourism following the designation has had a greatly transformative impact on local daily life, cultural practice, social relationship, and communities' relationship with the outside world. Along with the heritage designation and the development of heritage tourism, tulou and Hongkeng Village are converted into displays, performance stages, and commercial space. Tulou residents' relationship with and attachments to their home is under transformation. Now tulou are celebrated not only for their architectural value but also for the value of expressing cultural difference and local particularity while concurrently being embraced by humankind as a kind of global common property. While tulou residents are experiencing more frequent cross-cultural and intergroup encounters and interactions through the heritage process, they have become much more conscious and mindful of how they present and represent their culture and tradition. They actively curate their living space and recognize new forms of cultural expression.

Heritage tourism not only leads to the dramatic reshaping of village and residential space but also generates new forms of social interaction and power relationships. Chapter 6 examines how community members strategically re-negotiate everyday economics and new political power relationships in the process of socioeconomic transformation and modernization. As they are facing further complexity brought by the heritage process and the various agents involved in heritage policymaking and practice, they need to find ways to navigate power relationships, property ownership, and the politics of heritage. And the blurring of formerly distinct realities—private and public domains, local and global publics, "authentic" and performed ways of life—accelerates and intensifies dynamic interaction among the involved agents. The different actors may become involved with tensions, conflicts, negotiations, and compromises. Facing the new situation, local people have to adjust their positions and adopt effective strategies to negotiate living spaces, social relationships, and power balances. I conclude the book with a discussion on the concept and theoretical implication of negotiation in studying and understanding the intersubjective relationship of the macro structure and the intimate world of community members within the heritage process.

NOTES

1. The villager committee is a committee that usually constitutes three to seven village cadres who are selected or elected as local administrative leaders in Chinese villages. It is designed as a form of villager self-governance in the countryside. The functions of a villager committee include supporting the village development, administering the affairs concerning the land and other property owned collectively by the villagers, managing the public affairs and public welfare undertakings in a village, publicizing laws and policies issued by upper-level governments, conveying local opinions to upper-level governments, and mediating disputes among the villagers. The village cadres are oftentimes more resourceful compared to other community members.

2. The decorated archway is called *paifang* or *pailou* in Chinese. It is usually a tall gateway in a recognizable architectural style made of wood, stone, or the combination of both materials. The wooden paifang is typically constituted of two to six pillars, multitiered tile roofs, and colorfully painted beams with delicate sculptures. Those made of stone are usually plainer in shape and decoration. Paifang is traditionally used as a landmark for the entrance or border of a specific location,

such as a temple, a street, or a village. For more information on paifang, see Zhao Di赵迪, 2021, "Pailou mantan牌楼漫谈 [On Pailou]," *Minyi*, no. 4: 81–86.

3. My use of *symbol* here refers to an object, image, or sign that is used to represent or stand for something else. A specific example is that tulou architecture nowadays becomes a symbol that represents Hakka ethnicity.

4. In the noticeable national interests in safeguarding an internationally and nationally recognized/legitimated ICH, massive endeavors have been undertaken in the study, nomination, preservation, promotion, and commodification of ICH in cultural, political, and economic spheres across China. The prevalence of the ICH safeguarding in China is reflected in the statistics from the Ministry of Culture and Tourism. By the end of 2018, there were 40 ICH items from China inscribed in the UNESCO ICH Lists. Across China, 3,154 institutes had been established to work on national ICH safeguarding. In addition to the many national laws, regulations, policies, and sustaining and reviving plans, there were regional safeguarding regulations and policies. Within the state, there were 1,372 inscribed items on the national ICH list and 15,777 inscribed items on the provincial ICH lists. The number of national representative ICH inheritors (specific individuals holding "master" designations and possessing formal teaching duties) reached 3,068, and the number of provincial representative ICH inheritors reached 16,432. In terms of funding, in addition to the RMB 5.4 billion (USD 759 million) in national funding, there was RMB 4.6 billion (USD 646 million) from local governments invested in ICH safeguarding in the five years from 2013 to 2018. Regarding institutional programs that focus on the education, transmission, revitalization, and documentation of ICH, there were more than 670 ICH training programs for ICH practitioners organized by more than 110 higher education institutes, which attracted more than 28,000 participants. The "Revival of Chinese Traditional Arts and Handcrafts" program established fifteen working stations that supported the development of 383 ICH items with an effort to integrate them in "contemporary daily life." The ICH Documentation Project documented 1,114 national representative ICH inheritors who were sixty-five years old and over. Such statistics show the extraordinary degree to which people and resources are being mobilized for heritage-related practices, including developing legal, technical, administrative, financial, and educational measures for the safeguarding of ICH.

5. New countryside construction in China is one of the current nationwide macroeconomic development strategies aiming at sustainable development in rural areas in the context of fast urbanization. The measures and policies include increasing structural and financial support for rural areas, improving the living environment, enhancing support for the modern agricultural development, improving the infrastructure in rural communities, providing vocational training for rural community members, and facilitating increased living income. Culturally, it

emphasizes the preservation of traditional architecture and villages that represent local and ethnic historical and cultural features and values.

6. The four levels are a simplified description of the administrative divisions of China. In general, there are five administrative divisions under the central government: provincial level, prefectural level, county level, township level, and village level. But more specifically, in addition to provinces, the first tier of administrative divisions also includes autonomous regions, municipalities, and special administrative regions. There are twenty-two provinces, five autonomous regions, four municipalities, and two special administrative regions in China. The autonomous region is the highest level of minority autonomous entity with a comparably higher population of a particular minority ethnic group. A province or an autonomous region is subdivided into autonomous prefectures, counties, autonomous counties, and cities. And a county or an autonomous county is subdivided into townships, ethnic townships, and towns. Municipalities directly under the central government and large cities are subdivided into districts and counties.

ONE

—∿—

LOCALNESS IN THE TIME OF GREAT TRANSFORMATION

TULOU, AS A PARTICULAR vernacular architecture style that is found in a confined area of China, is an integrated part of the local communities where it appears. In the past, the local geographical features, historical development, and sociocultural dynamics of the places where these sites are located are significant formative factors of tulou as a local form of architecture. Now, the cultural richness of the local community—and particularly of tulou—have been directly targeted in the more recent construction of localness by heritage tourism. The landscape, history, ethnicity, symbols, and meanings within "the most intimate life-worlds" of local community members are all included in the local strategic plan for development and subjected to touristic commodification (Marcus and Fisher 1999, 39).

UNESCO's designation of tulou constructions as World Heritage, and the tourism development that followed it, opened these once isolated places to the world and has been a driving force for the local pursuit of modernization. The development of tourism has diminished the traditional agricultural lifestyle while also contributing to the revival and invention of other local traditions associated with an agricultural society. Moreover, it has turned convenient and potentially appealing sites for tourism development, such as Hongkeng Village, into enclosed, curated, and highly regulated spaces to be consumed by tourists.

Yongding County, like other parts of China, is going through a period of rapid social transition. In the process of accelerated economic and tourism development, the imposition of political and cultural authority (as demonstrated in the discussion of land acquisition, relocation, and policymaking in this chapter) often results in tensions between outside forces and local autonomy, modern development and continuity of tradition, and governmental ambition and community interests, which require the various agents to engage in constant adaptation, navigation, and negotiation. These themes are the reoccurring focus of this book, just as they are prominent in the contemporary life experiences of those who reside in this small corner of China and the Hakka world.

YONGDING COUNTY IN THE HISTORICAL PAST: A REMOTE AREA WITH LIMITED RESOURCES

The World Heritage Site Hongkeng Village is one dot in a connected network and should be examined in its broader context. The village is in Hukeng Township, Yongding County, which is located in the southwest of Fujian Province under the jurisdiction of Longyan prefecture-level city, bordering the predominantly Hakka counties of Dabu and Meixian in Guangdong Province to the southwest and the predominantly Minnan counties Nanjing and Pinghe in Zhangzhou Prefecture to the east and southeast (map 1.1). The prefecture city of Longyan is to the northeast, and another predominantly Hakka county in the prefecture, Shanghang, is to the northwest. Hakka consists of more than 99 percent of the population of approximately four hundred thousand people in Yongding County, and the Hakka language is the local dialect spoken by Hakka people in this region. The county is divided into twenty-three township-level units, under which are villages. The county's political, executive, and cultural center is in the county seat Fengcheng Town. The county's Yongding River, Tingjiang River, and Jinfeng River are filled with abundant water brought by frequent rainfalls in this subtropical area, as are the major rivers that connect the twenty-four towns and townships scattered throughout the valleys formed from the chain of undulating hills sloping down from the northeast toward the southwest.

The county name Yongding literally means "peace forever." Narratives from local documents traces the origin of the name to a turbulent time in the region's history. The area of what we now call Yongding was under the governance of Shanghang County during the Tang dynasty (618–907). During the Ming dynasty (1368–1644), there were rebellions along with robberies and fierce fighting in the region. The governor of the county reported to the central government that it was difficult to govern the area given that it is "remote and savage."[1] He proposed that a separate county government be set up to enhance local security. Thus in 1478, the Fujian provincial government sent a specialist to

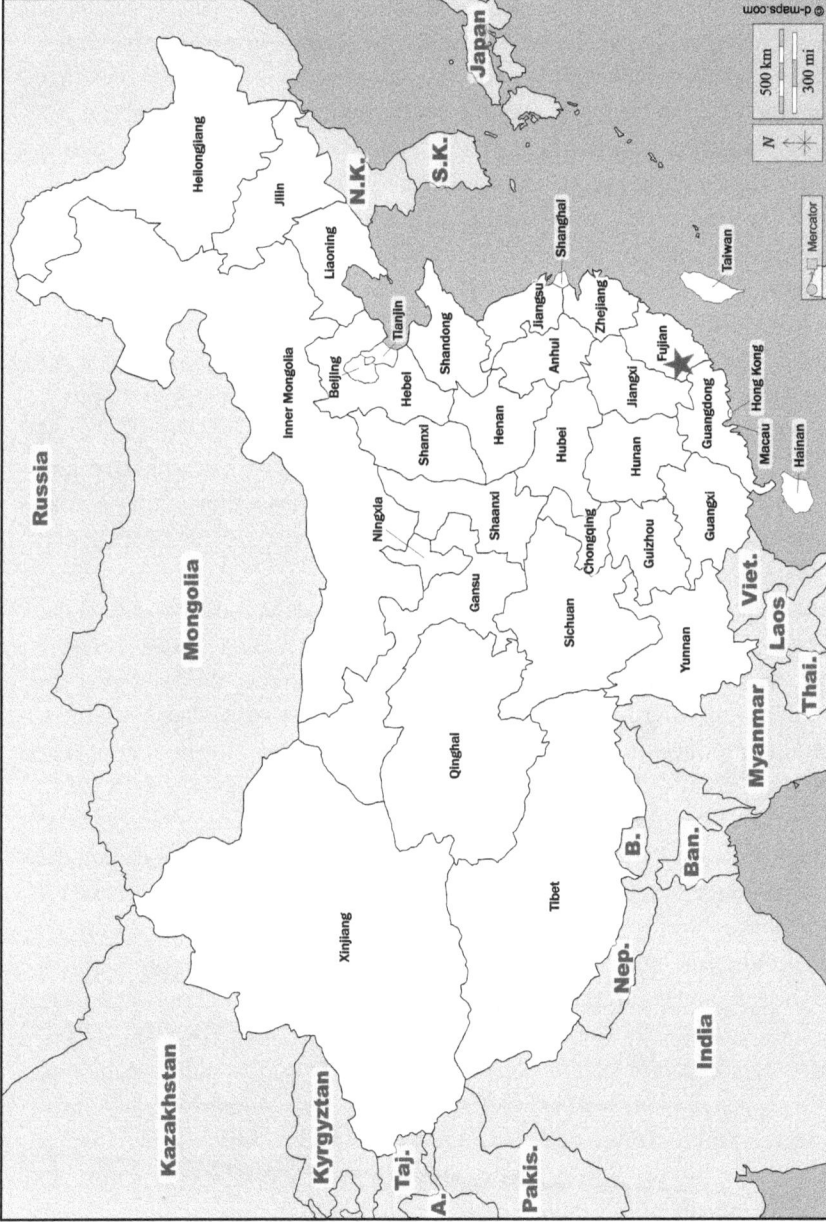

Map 1.1 Location of Yongding County, Longyan prefecture-level city, Fujian Province, China. © Daniel Dalet, https://d-maps.com/carte.php?num_car=11572&lang=en.

map and mark the land as well as choose a site for the new county government center. With good wishes, the county was named Yongding.[2]

Even when the county was under independent governance, historical documents such as local gazetteers often describe the people in the area, including both the local non-Hakka groups and the descendants of Hakka groups that migrated to the area from places in the north to Fujian Province, as a "barbarous and violent" people living in a harsh natural environment of precipitous and dangerous mountains. This harshness is partly related to the geographical features. Approximately 80 percent of the land in Yongding County is hills and mountains, and less than 10 percent of the land can be cultivated in the narrow valleys. Terraced farming is a common form of agriculture, with rice as the main crop in the region. Some highland fields offer only a single harvest because of the cooler water and earlier frosts found at the higher altitude. So people make full use of the lower land that can yield two crops per year. A small amount of sticky rice is grown primarily for making rice wine and local snacks. Other crops cultivated in the area include millet, beans, barley, and sweet potato, which are now only occasionally consumed although they played an important role as the staple food for the people during difficult times in history, when rice was in short supply.

Economic resources are limited for most rural residents, and for quite a long time, tobacco has been the single most important cash crop in the area. The cultivation history of tobacco in the area dates back to the late sixteenth century, during the Ming dynasty—in other words, when it was still new to China. The county produces high-quality tobacco, which earned it the honor of being dubbed the "home of tobacco" (*kaoyan zhi xiang*, 烤烟之乡) in the early twentieth century. From 1988 to 2000, more than one-fifth of the cultivated land was used for tobacco cultivation with the portion rising to almost 50 percent in the peak years.[3] In recent years, the significance of tobacco has declined dramatically with the development of other forms of economy and income sources, including tourism. However, the county maintains its place on the national list of the forty-one "excellent tobacco cultivation bases" in China, and many top brand cigarettes in the country contain tobacco from Yongding. Other historically significant income sources for local people include the making of candles, paper, plant seed oil, tea, bamboo products, and other handcrafts. The most notable natural resource in the county is coal, which under current circumstances has provided a small group of people with the opportunity to become financially privileged.

RECENT DEVELOPMENT AND SOCIAL CHANGES

Since the implementation of China's reform and opening-up policy in 1978, Yongding County, as with other parts of China, has experienced dramatic transitions in political, economic, and cultural spheres. During the last three decades especially, Yongding County, which was once considered a remote and underdeveloped place, has had more contact with the outside world than at any other time in its history in the process of pursuing industrialization and modernization. In the process, governmental strategic planning and policies not only play a vital role in local economic development but also have a significant impact on community culture and personal life. The profound effects of state power and social change continue to be felt in everyday life even as economic pursuit becomes fundamental to Chinese—and Yongding—lifeways, as illustrated by the construction of the Mianhuatan Hydroelectric Power Station as a grand project in the county.

The rapidly growing economy increases the demand for electric power in the region. Taking advantage of the abundant precipitation and mountainous terrain, the local government invests heavily in harnessing hydroelectric resources. In addition to many small hydro plants, in 1998, the government started to construct Mianhuatan Hydropower Plant, one of the biggest in this Southeast China province. It took three years to finish the plant.

While the plant has made a magnificent contribution to further economic growth in Yongding County and neighboring regions with its capacity to generate 1.52 billion kilowatt hours of electricity a year, the project has also had a significant impact on local communities and local residents' lives. More than five thousand acres of farmland were flooded and more than thirty thousand people were relocated because of the massive project. My friend Luchun's family, whom I stayed with during my visit to the county seat Fengcheng Town, is among those relocated to a recently built allocation site in a suburb of Fengcheng Town during the construction of the plant. The rows of concrete buildings centered on a small garden were constructed by the government, and villagers were asked to draw numbers written on pieces of paper to decide the distribution of the houses. Those who used to live in the same building with a big family or lineage branch now live separately in single-family units. Luchun's mother, who used to be a farmer, is now a temporary worker. During my stay, she got up at five in the morning and went to work at the highway construction site. Infrastructure construction is among other reasons that some farmers leave the land they lived on for their entire life and relocate to a different place. In the wake of such dislocations and disruptions, dramatic changes may happen to their social relationships and lifestyles (Ma 2021; Shi and Wei 2021). In the case of Mianhuatan Hydropower Plant, those who used to live as a small community under the same roof in a rural village now live separately in a new community in the suburb of Fengcheng Town.

The hydropower plant is only an example of the Yongding County government's efforts toward industrialization and modernization. Since 2011, the county government has been implementing a strategic developmental plan called the "dual-core, dual-axis, multi-location" plan (fig. 1.1). Specifically, "dual-core" refers to industry and tourism, which are targeted by the local government as the two major forces for economic growth. "Dual-axis" means two sets of transportation lines that run through the county and converge at the county center. One axis is the railway, state road, and highway from the north to the south connecting the prefecture-level city Longyan in the north, the industrial center in the northern part of Yongding County; the city Fengcheng Town, where the county seat is located; and the wealthy neighboring province Guangdong in the southwest. The other axis is the recently developed tourist highway from west to east that links such scenic spots as the tulou World Heritage Sites in Hukeng Town. Transportation is applied as an instrument that accelerates the flow of people and resources both inside and outside the county, so as to boost economic growth for cities and towns along the traffic lines. Other major cities and towns that are not on the major traffic axes are referred to as "multi locations." Governmental resources and support services focus on these places to boost their development.

Lack of infrastructure, such as transportation, used to be a significant barrier to the development of the region. This mountainous, isolated, remote area was hard for outsiders to reach. Local people used to get in touch with the outside world through very narrow and rugged paths through hills and mountains. To attract more tourists and investors to the area, the government aimed to build a highway network that connected the area to the outside world in a fast way. During the period of my ethnographic research in 2011 and 2012, there were three key highways under construction: Guwu Highway, Yonghe Highway, and Shuanyong Highway (fig. 1.1). When Luchun's uncle picked me up and drove me to Luchun's house from the county seat bus station, he told me that now it only takes about an hour to get to Longyan by highway. This highway is the S67 Highway, which was finished in 2011, except for a section in Hulei Township due to some local residents' refusal to relocate because of disagreement over the land acquisition fee; if the government could successfully negotiate

Figure 1.1 Yongding County strategic development plan. (Figure created by author based on materials and information provided by Yongding County Bureau of Urban and Rural Construction and Planning)

with them to complete the route, the travel time could be further shortened. As with the construction of a massive hydropower plant, accompanying highway construction are land requisition and potential relocation of village residents, which may result in loss of farmland, loss of sites for common social memories, a change of lifestyle, and new encounters between relocated residents and those who have been living in the relocation sites.

In addition to the "dual-core, dual-axis, multi-location" plan, the county government established five types of development paths for different sections of the county based on the location and major resources that they possess as assessed by the local government. For instance, the development model of the northeastern part of the county is set as an industrial type. Fengcheng

Town, as the county seat, is targeted as a multifunctional city type. Hukeng and Xiayang, where a large number of the UNESCO-designated vernacular architecture tulou are located, is set to focus on developing tourism. According to a document released by Yongding County Bureau of Urban and Rural Construction and Planning, four towns that are rich in touristic resources (in this case "touristic resource" means a considerable number of "representative" tulou and noteworthy natural scenery)—Xiayang, Fengshi, Hukeng, and Longtan—are marked as "central towns" in the development plan. Among them, Hukeng occupies a special place in the government's ambition for local development. As one of the most important tulou tourist sites and a designated World Heritage Site, it is targeted as "the characteristic tourism-oriented town and provincial historic town" that should "synthesize and use its culture and history as touristic resource[s]."[4]

One of the goals of Fengcheng Town is to construct a "scenic and livable tourism city." The service industry, especially service in tourism, is set as the main impetus for economic development of the city. Fengcheng used to be an ancient city full of historical sites and traditional houses. However, now what people see is a fast-developing city with significant ongoing construction. When I first visited the county seat, I was greeted by a grand marble gate with the name of the city in a golden color that marked the entrance of the county center. Right next to the gate is a river running through the city. The south side of the river was filled with real estate business offices in modern architectural style in front of shining new tall concrete residential buildings. The north side of the river is the old town, which is not recently developed like the south bank of the river. Instead, the streets are lined with two- to four-story buildings painted beige, resembling the locally iconic color of the tulou walls. When carefully navigating through cars, motorcycles, and pedestrians, I saw a large map on the fence of a construction site that laid out the government's plan for constructing an "ancient town." The so-called ancient town was a totally new construction that was designed to be located in the suburban portion of the city, so as to better attract tourists.

The value and ideological trends that perceive countryside life as backward and urban life as the symbol of modernization contribute greatly to the local desire for, and pursuit of, urbanization. Meanwhile, in the context of tourism-oriented development, traditional and historical qualities are valued for achieving the goal of local economic development and modernization. Despite the local government's efforts to promote historical sites and even invent an ancient town, Fengcheng Town expressed an eager and ambitious goal for urbanization and modernization. The county blueprint divides the period from 2011 to

2030 into three stages for urbanization. By 2015, Fengcheng Town expected 112,000 permanent urban citizens, reaching an urbanization rate of 90 percent. By 2020, the city and its surrounding area aimed to have 137,000 permanent urban residents with an urbanization rate of 95 percent. It sets a goal to have 99 percent of its population of 200,000 live in an urban area by 2030. With the growth of the urban population, the county and city plan to construct more "modern communities."[5]

Along with the government's goal for urbanization, another noticeable social movement that has had a transformative effect on local communities is the construction of "the new socialist countryside," a campaign proposed by the Chinese government in 2005. This was a governmental effort to promote rural development and narrow the developmental gap between urban and rural areas. While planning their work for the second half of the year 2012, the Yongding County Land and Resource Bureau pointed out that, in line with the construction of the new socialist countryside, it planned to approve more land for the use of construction. According to public information posted by the Yongding County Land and Resource Bureau, it "urge[s] villagers in rural areas to tear down old houses and relocate to new communities constructed under the design and planning of local government so as to greatly improve living conditions in rural areas." Once again, change of tradition, social network, and local daily life resulting from the acquisition of land; demolition of houses and landmark sites; and relocation are issues that require close examination in this context.

The projects, movements, and development plans illustrate how the developmental path planned and crafted by the local governments can impact and change local people's living environments, social paths, and daily lives. Furthermore, the recent policy that allows farmers and other individuals to subcontract or sell and buy land has significantly transformed modes of agricultural production and local lifestyles. According to statistics released by the Yongding County Land and Resource Bureau, during the first three quarters of 2011, 39.5 percent of farmland had been transferred to new owners.[6] Concentration of farmland ownership resulted in more efficient and profitable large-scale plantations. At the same time, many farmers who used to grow food for their own daily consumption no longer had land. Furthermore, many of those who still own farming land stopped agricultural activities because the time and labor required for farming is not as lucrative as city jobs. A large number of farmers and young people from the rural area migrate to work in the cities, which is one major cause for the related phenomena of "empty nesters" (*kongchao laoren*, 空巢老人) and "unattended children" (*liushou ertong*, 留守儿童), in which old people and children are left behind in villages when rural laborers between the ages of twenty and sixty leave home as migrant workers. Some scholars also refer to this kind of village as a "hollow village" (*kongxin cun*, 空心村). This is a commonly observed phenomenon in rural China that has been widely reported by media and discussed by social scholars (Liu and Liu 2010; Liu and Yang 2013; Xue 2001; Yang and Yao 2021; Zhao 2021). In many parts of rural China, agriculture has lost its dominance in people's daily economy and everyday life. In this larger context, I examine the Fujian Tulou World Heritage Site Hongkeng Village as an epitome of Chinese rural society in the wake of rapid social change that is often hastened by state power and governmental development strategy and ideology.

HONGKENG VILLAGE

Hongkeng Village, one of the most famous tulou "scenic spots," was my major field site in Yongding County. In 1993, it was given the title Yongding Hakka Tulou Folk Culture Village, indicating that the village was selected to represent the region with its Hakka ethnicity, tulou, and folk culture. With more than forty tulou of various shapes located collectively in the village, Hongkeng is registered by UNESCO as a World Heritage Site known as the Hongkeng Tulou Cluster (fig. 1.2). Hongkeng is under the township of Hukeng in Yongding County.[7] As stated previously, Hukeng Township is positioned by the county government as a historic town with the main purpose of focusing on developing tourism. The major tourist attraction in Hukeng Township is Hongkeng Village as both a traditional Hakka community and a UNESCO World Heritage Site.

Hukeng refers to both the township and the village where the local administrative and market center is located. It is only a five-minute drive from Hongkeng to the town center Hukeng, where the local periodic market attracts people from neighboring villages to attend the "market days" every five days. People take the bus, ride motorcycles, or walk to the market to sell farm products and handicrafts or buy daily necessities. Now there is also a small-size morning market in the center of Hongkeng Village where people can sell and buy meat, vegetables, fruits, and homemade rice snacks (fig. 1.3). People who own a car also go shopping farther away, in places such as Fengcheng Town and Longyan.

The increase in car ownership and the newly constructed roads have greatly improved local people's mobility. However, few people were able to afford to buy a car before the development of the heritage touristic economy. And the road that connects the town market center to Hongkeng was not paved until

Figure 1.2 Layout of Hongkeng Village with the tulou buildings along the Jinfeng River. (Figure created by author based on field observation, consultation with villagers, and materials provided by the Archive of Yongding Office of Cultural Heritage)

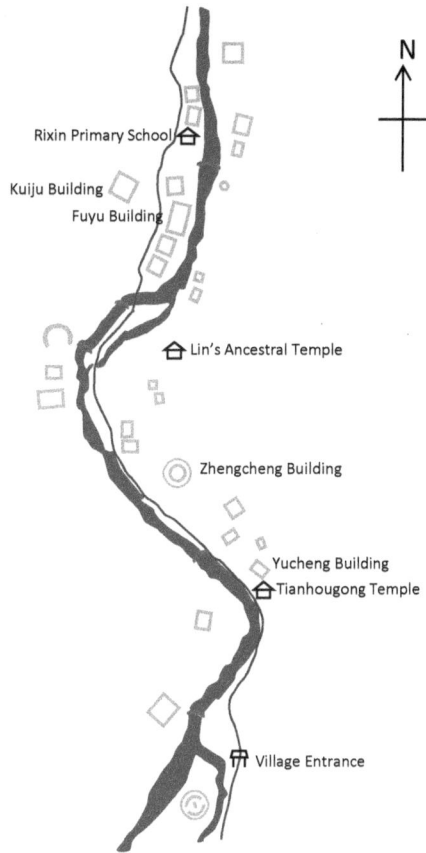

the early 1990s. To the local people, Xiamen is the nearest cosmopolitan city—about two hours' drive from Hongkeng on the recently constructed highway. Longyan and the county political center Fengcheng Town are smaller cities that are relatively easy to reach. Although the younger generation became familiar with Xiamen and other cities as students and city workers, many members of the older generations have never traveled out of the township.

The narrative and image of remoteness discussed previously are reflected in the Hongkeng villagers' description of their living environment. In the local Hakka language, people often use the word *shan* (山) to describe the spatial and social correlation between different places. Shan literally means "mountain" or "mountainous." In the local language, the word not only describes the geographical feature of a place but also refers to the socioeconomic status of a place

Figure 1.3 People selling and buying food at the morning market in Hongkeng Village. (Photo by author)

in general or in comparison to other places. Usually, shan implies that a place is socially isolated and economically underdeveloped. For instance, compared to Fengcheng Town, Hongkeng villagers perceive their village as shan—mountainous, remote, and less developed. When talking about places farther away from the town center and deeper in the mountains, Hongkeng people do not think of their village as being as much shan as those other places. Also, the shan image and status can be changed with improved accessibility and economic development. As one of the residents in Yucheng Building said, Hongkeng is not as shan as it used to be thanks to the local development of tourism related to the designation of the village as a World Heritage Site.

After climbing up the hill on the north side of the village along the road paved with slabs of granite, visitors reach the brand-new sightseeing pavilion on the top, with views of the whole village and the undulating mountains surrounding it (fig. 1.4). Hongkeng Village is surrounded by hills and mountains on the east, west, and south. The Jinfeng River runs through the village from

Figure 1.4 A view of Hongkeng Village from a sightseeing pavilion on the top of a hill on the northern end of the village. (Photo by author)

north to south. Bamboo and plantains adorn the banks of the river. Tulou, in various sizes and shapes, and patches of fields lie along the relatively flat land along the banks of the river, shaping the village that spreads from the north to the south in a long narrow space.

Among the many buildings with beige walls and black roofs, the two structures that stand out in the view of the village from higher up are Zhencheng Building, a massive round structure in the center of the village, and the Lin Family's Ancestral Temple (*Linshi jiamiao*, 林氏家庙), a symmetric traditional architectural structure with tall limestone monumental pillars to commemorate some of the outstanding ancestors of the Lin lineage in the front (fig. 1.5). The two structures are landmarks of the village with different social and cultural significances.

The Lin Family's Ancestral Temple represents the social relationship based on the Hakka lineage system. The villagers in Hongkeng are the descendants of the same ancestor, with Lin as their family name. In essence, all of the villagers

Figure 1.5 The Lin Family's Ancestral Temple with twenty-four pillars in front of it. (Photo by author)

belong to a big Lin family. The Lin Family's Ancestral Temple is an inward-oriented space that, in physical form, reifies the kinship relationship between the villagers as the descendants of the same lineage. It is located at the foot of a small hill in the center of the village—an auspicious site said to be carefully chosen by the villagers' ancestors. It is a spiritual and sacred space for ancestral worship and ceremonies in the village. According to local narratives, Yong-song, a member of the sixth generation of the Lin family in Hongkeng, built the temple in memory of their ancestors in 1483. In local narrative, Hongkeng became prosperous due to the thriving business of producing tobacco cutters and tobacco in the late nineteenth and early twentieth centuries. During that time, many people accumulated wealth to build the enormous houses known as tulou, including the most famous ring-shaped Zhencheng Building. Some villagers donated money to expand and modify the temple in 1923. At that time, there were memorial pillars in front of the temple to honor villagers with high cultural or political merit. To the villagers' regret, those pillars were destroyed during the Cultural Revolution. After the village was nominated as a World Heritage Site, twenty-four new pillars were reintroduced.

ANCESTRY AND FOLK NARRATIVE

The sense of community bond is strong among Hongkeng villagers, largely thanks to their shared history and the social network of their lineage system. Stories about their common ancestors have been widely circulated in the village. Local genealogy traces the Lin family's origin back to the Shang dynasty (1600–1046 BCE). Denghui, a retired teacher of Hongkeng Village Elementary School who is regarded by the community members as the most knowledgeable person about the village history and is in charge of the reediting work of *Lin's Genealogy*, told me a legend about the origin of the family name Lin:

> In the Shang dynasty period, the ancient patriotic Prime Minister Bigan was killed by the tyrant. His wife escaped into the woods and gave birth to their child. Lin, which literally means woods, was given to the posthumous child by the emperor of the new dynasty Zhou in honor of his father. Thus, Bigan and his son are regarded as the earliest ancestors in the Lin's genealogy. During the Xijin dynasty, warfare forced many Han Chinese to move to the more remote regions in the south from Central China. Lin's ancestors were among those who migrated to Fujian Province. They first settled on the border of western Fujian. Later on, the Lin's descendants developed very quickly and could be found all over the province.

As shown, the narrative of the history of the Lin's family and village is largely about their migration and settlement as well as patrilineal development and grouping. Such narratives are commonly found in written genealogies and gazetteers. Denghui continued with stories about the origin, migration, ups and downs, and settlement of the Lin's ancestors:

> During the chaos caused by wars in the time of the Xijin dynasty, the forty-seventh generation in the Lin's genealogy Lu served in the military and moved to the Yangzi River delta with the army. He later distinguished himself in the army and was promoted to general. After the new emperor was enthroned, Lu was appointed governor of Fujian Province. And he was the first ancestor of all Lin's branches in Fujian.

Wende, the thirty-eighth generation of Lu's descendants, moved to Ninghua County (which is widely recognized as the first settlement of Hakka in Fujian) because of warfare. At the end of the Song dynasty, Wende's sons were forced to move to more remote areas after some fights with local gangsters. Balang, the eighth of nine brothers, moved to Baisha Town in Shanghang County. His great-grandson Daxing had six sons. The eldest one, Maoqing, was born in the 1320s. Maoqing had two wives and the second wife, Zheng, bore two sons. Maoqing's wives and sons had to leave their hometown because his father was accused of building a private fortress. Maoqing arrived at Fushi Township in Yongding County with his second wife and second son. He settled there and married his third wife, Zang. They had four sons.

Thus, Maoqing is our earliest ancestor to settle in Yongding. After Maoqing passed away, his wives and sons built an ancestral temple to honor him. A local gentleman's ignorant stableman tied his horse to the niche in our ancestral temple. The horse jumped and tore down the memorial tablet in the niche. The stableman, after being strongly rebuked by his master for his negligence, committed suicide in our ancestral temple. The incident resulted in conflicts between the Lins' clan and the stableman's clan. In order to avoid hatred between clans and to stay out of trouble, Maoqing's wives had to leave their home. His second wife, Zheng, and her sons settled in Xiazhai and Liantang. His third wife, Zang, and her sons moved to Yukeng, Yantai, and Hongkeng.

It was in 1387 when Zang and one of her children, Sansanlang, arrived at Hongkeng and settled here. Another son settled in Yantai. During festivals, both her children and grandchildren in Hongkeng and Yantai wanted to celebrate with her. So she came up with a solution: Hongkeng would celebrate the Dragon Boat Festival and Double Ninth Festival a day before the actual date of the festival, and Yantai would celebrate the Lantern Festival and the Mid-Autumn Festival a day before the actual date of the festival. The tradition is carried on even now.

This narrative circulates in the village, mainly in the form of written genealogy but also in oral form. In the story, the migration path of the villagers' ancestors is clearly traced and narrated. It is a shared memory and narrative of the village that enhances the social identification of Hongkeng villagers.

As demonstrated in chapter 4, this kind of narrative also explains the social organization and grouping in Hongkeng Village. Now Sansanlang's descendants have developed five lineage branches in Hongkeng. They cooperate with each other through the lineage system while also competing at times for limited resources.

HARSH LIVING ENVIRONMENT, MIGRATION, AND OVERSEAS CONNECTION

Echoing the narratives about Yongding County being a remote and underdeveloped region, many of my informants in Hongkeng Village carry the same narrative for life in Hongkeng that is grounded in the past. The pioneer settlers in Hongkeng, like other Hakka who migrated to the mountainous region in Southeast China, had to create their own farmlands through terracing and other techniques for lack of fertile land. Rituals were performed to ask for blessings of a good harvest from the earth god. In Hongkeng Village, the ridges of some terraces were carefully constructed with stones to protect them from flooding and landslides. Still, the poor and limited land could not support the growing population in the village. Many had to find other ways to make a living. A few people whose families could afford the tuition distinguished themselves through formal education. Many others became apprentices of craftspeople and other occupational skills such as barbers, carpenters, and housebuilders.

Some people tried their luck overseas. The majority of them traveled to Southeast Asia. These people not only suffered from the pain of leaving their family and friends at home but also risked their lives in the long journey on the dangerous ocean and struggled to survive in foreign lands among peoples of different ethnicities and nationalities. Many Hakka folk songs express the feelings of those who had to say goodbye to their families, lovers, and friends under such circumstances. A song that is most well known and now often sung in touristic performances is titled "A'ge Is Leaving Home for Southeast Asia" (*A'ge chumen xia Nanyang*, 阿哥出门下南洋), expressing the farewell words of a wife when her husband (A'ge, which literarily means "brother" but in Hakka folk songs is often used to refer to a husband or a lover) departs to Southeast Asia:

A'ge is leaving home for Southeast Asia
To a foreign land across the ocean
Please take good care of yourself
So you will return home healthy and well
with success and honor

A'ge is leaving home for Southeast Asia
It is a long journey ahead
Upon your arrival at the destination
Please send letters back home immediately
So that I won't be so worried about you

Brother is leaving home for Southeast Asia
Our love will last despite the distance
I will take care of our parents,
Please count on me for things at home
Please have no worries in the foreign land

A'ge is leaving home for Southeast Asia
I have a few farewell wishes
Send money back home
As we have old parents to support
Hope you will be flushed with success in the foreign land

A'ge is leaving home for Southeast Asia
I have a few questions to ask
When will you be back for home?
At the dock I will be waiting for you
The reunion will be full of happiness

This is one version of the traditional Hakka folk song that was regularly sung by the local performer Li Huifeng at Qingcheng Building in Hongkeng Village as part of the tourism performance when I visited it in 2012. The song expresses the feelings and worries, expectations and hopes, values and morals of those who saw off family members and loved ones to foreign lands and waited for their return. Many of them settled overseas and rarely returned home. The great-grandfather of my informant Ziwu (who, as detailed in the introduction, also introduced me to some of my hosts during my stay) was one of those who didn't come back, leaving his wife and children in Hongkeng Village. Talking about those faraway family members and relatives often arouses a kind of sentiment that mixes familiarity and exoticness. Local people call those immigrants *fanke* (番客), which literally means "foreign guests." As expressed in the folk song, fanke used to be the only hope of some families in the region. If they were lucky and worked hard enough, they might not only provide their families with financial support but also glorify the family for being "successful." One important financial resource for the construction of tulou was said to be the money sent back by fanke.

Figure 1.6 Community member burning incense at Tianhougong Temple during the Dragon Boat Festival. (Photo by author)

As more and more people in Hongkeng went overseas, Hongkeng villagers built Tianhougong Temple as a sacred space to pray for the safety and well-being of their relatives who were far away from home. The major deity in Tian-hougong Temple is Matsu, the goddess of the sea who protects those fishing or traveling on the ocean. Matsu is especially popular in the coastal areas in the southeast and south of China as well as in Chinese communities in Southeast Asia. The legendary figure Matsu, whose original name is Lin Mo, shares the family name, Lin, with people in Hongkeng, so the villagers also call Matsu "grandaunt" (*gupo*, 姑婆) and call Tianhougong Temple "grandaunt's temple" (*gupo miao*, 姑婆庙). The ocean deity is venerated inland and thus creates a closer connection between Hongkeng Village and those who live near the ocean and worship Matsu.

Today, Tianhougong Temple is not only a sacred space of the community but also a landmark and a tour site in the village (fig. 1.6). It is constructed of bricks and carved or painted wood. On the roof beam with upturned eaves, there are two painted clay dragons. Along with the Lin's Ancestral Temple, Tianhougong Temple is another important sacred space for local spiritual life. On important occasions such as festivals, marriages, childbirths, and college entrance exams, Hongkeng villagers pray in the temple for blessings. On the first and fifteenth days of each lunar month, some senior villagers perform the ritual of offering sacrifice and burning incense at the temple. After the development of heritage tourism in the village, Tianhougong Temple has also become a touristic space that represents the historical background and current religious practices of the village. It is registered as a "Site of Historical Interest and Cultural Relics at County Level" by the local government. The temple not only connects Hongkeng Village to the overseas Hakka population but also arouses a sense of identification and connection among tourists from Guangdong, Hong Kong, Taiwan, Macau, and other places where Matsu is also commonly worshiped.

ECONOMY AND SOCIAL DEVELOPMENT IN HONGKENG

As in other parts of Yongding County, tobacco used to be the major cash crop for Hongkeng villagers. With the growing popularity of Yongding cut tobacco across the country, Hongkeng villagers started to make tobacco cutters, which later became the predominant industry in the village. The profit from tobacco and tobacco cutters were two of the major financial resources for the construction of tulou in Hongkeng. Entering the twentieth century, as told by Denghui, the business of Hongkeng handmade tobacco was beaten by Japanese competitors with their modern machines: "Famous brands in Hongkeng produced the best tobacco cutters at that time, which was a very profitable business. The brands opened stores in big cities like Shanghai. The storekeepers only needed to work for half a day each day and enjoyed the rest of the time for entertainment. Some Japanese saw the profit in the business and started to produce tobacco cutters with their advanced machines. The handmade workshops in Hongkeng could not compete with the assembly line work. Most workshops went bankrupt and a few reformed and made farming tools and furniture instead."

This account provides a glimpse of local craft industry prior to the founding of the people's Republic of China in 1949. In the late 1950s, the Communist Party of China started the so-called Great Leap Forward campaign, which introduced agricultural collectivization to rural areas. Peasants in Hongkeng had to join state-operated communes and stop private farming. As my informants Denghui and Ahen told me, the owners of large tulou were viewed by the government as landlords and were forced to give their houses to "the people."

Figure 1.7 Peeled persimmons being sun-dried on a bridge in Hongkeng Village. (Photo by author)

Villagers were grouped together as production teams and "occupied" the major tulou.

The collectivization of agriculture was brought to an end with the Chinese economic reform that began in in 1978, and farming land was redistributed to community members. Once again, the villagers were allowed to grow cash crops as they wished to. In addition to tobacco, people started to plant tea trees on the hills. There are still two tea-drying workshops that have machines for drying tea leaves. One of them is owned by Ziwu's family, some members of which share the tulou with my hosts in Yucheng Building. When the profit of selling tea went down, many Hongkeng villagers replaced the tea trees with fruit trees, mostly persimmons. Persimmons were harvested in the fall for making dried persimmons (fig. 1.7). In 2000, Ziwu's family built a workshop that produced more than six hundred pounds of dried persimmons each day. The dried persimmons were sold to businessmen from neighboring Guangdong Province and other places outside of the village. During my visits, the persimmon business had greatly declined. Only a few families still made dried persimmons, and at the time of my fieldwork, these were mainly sold to tourists in the village. After tulou were designated as World Heritage, tourism has become the dominant business and major source of income for many of the villagers. Ziwu's family no longer run the dried persimmon workshop. Instead, they make rice wine and sell the wine to tourists. The business was still being operated when I revisited Hongkeng in the summer of 2023, but it seemed to be downsized due to the impact of the COVID-19 pandemic and the decreasing number of tourists as a recently constructed highway led many tourists to another nearby tulou site. Ziwu's parents were too old to continue with the business and retired. Even though his sisters-in-law still helped with the selling of rice wine, his two brothers who used to run the family business with him started their own new careers. One became a construction contractor, and the other was elected as a member of the village cadre committee.

HAKKA ETHNICITY AND LOCALNESS

During my time in Hongkeng and Yongding, *Hakka* and *tulou* were the two words I encountered most frequently and heard almost every day. Hakka ethnicity, like tulou, has become a significant cultural, political, and economic resource for Yongding County and its surrounding areas. As Yongding focuses on developing tourism, Hakka ethnicity and tulou are conveniently used to promote a local image to outsiders. The otherness of ethnicity attracts curious outsiders to experience and explore the localness. Burgeoning ethnic tourist villages and theme parks have been a significant part of the tourism industry in China for more than two decades (Chio 2014; Liu and He 2021; Oakes 1998; Schein 2000). In the case of Yongding County, Hakka ethnicity along with the local vernacular architecture of tulou is promoted to attract national and international tourists. When I interviewed the director of the Yongding Gazetteer Committee, Liu Zelin, he emphasized that Hakka ethnicity and tulou are the two most recognizable features of Yongding County that combine to showcase this unique locality to the world. "Promoting locally characteristic features is the most likely way to make a place known internationally, which is very true in the case of Hakka ethnicity and tulou," he remarked.

The term *Hakka* (*Kejia* 客家), or *Hakka people* (*Kejia ren* 客家人), literally means "guests" or "people from the guest families." The majority of Hakka in mainland China live in concentrated communities in west and southwest Fujian, east and northeast Guangdong Province, and south Jiangxi Province

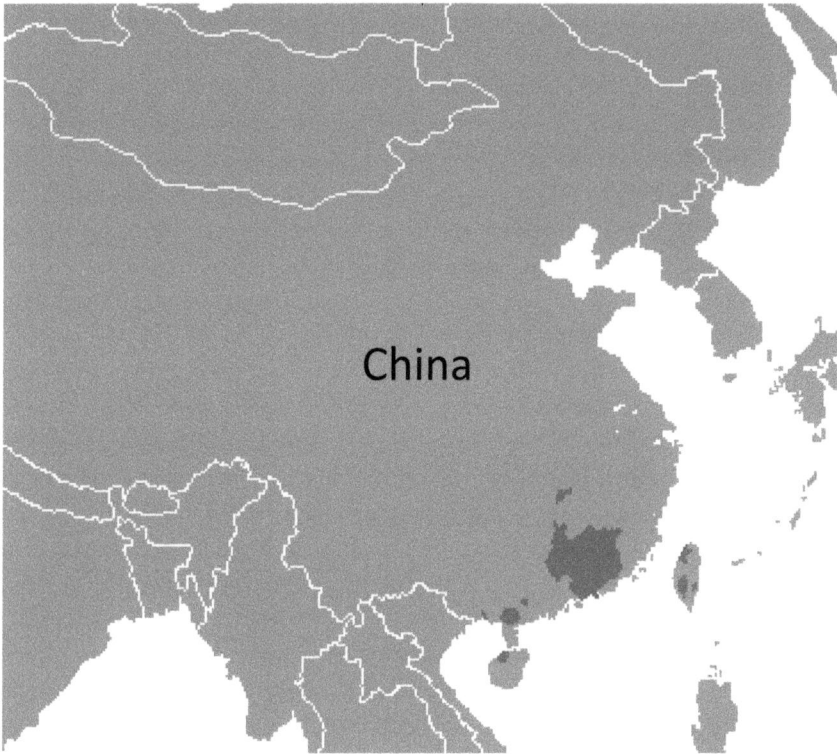

Map 1.2   Location of Hakka-concentrated communities. © Fobos92, CC BY-SA 4.0, https://creativecommons.org/licenses/by-sa/4.0, via Wikimedia Commons.

(map 1.2). There are also Hakka people living in Hunan, Guangxi, Sichuan, Hong Kong, and Taiwan as well as a large diaspora overseas. The name Hakka suggests that they are not the original inhabitants in the regions where they now live. The origin of Hakka is still an unsettled issue.[8] One of the most widely accepted suggestions is that Hakka people migrated from Central China during different times because of warfare and natural disasters (Hsieh 1929; Luo 1975). After settling in remote areas of Southeast and South China, they gradually adapted to the local environment. Eventually the "guest" from the north became dominant over the native groups in many of the Hakka-concentrated areas. They are often characterized as "a strong, hardy, energetic, fearless race" living in lineage-based villages with a strong sense of unity (Hsieh 1929).

Dwelling in the isolated mountains and interacting with the native groups, the Hakka people developed their own distinctive culture and language. Within official mainland Chinese designations of ethnicity, the group is identified as a subgroup of the majority Han. Interestingly, the awareness and celebration of Hakka ethnicity seems to be at peace with their officially stated Han identity despite their frequent conflicts with other Han groups and communities in history (Cohen 1968). According to my own experience as a Hakka and my observation of, and interviews with, the local Hakka people, despite the local promotion of the Hakka image, Hakka people are usually not politically defensive of their ethnic distinctiveness in Fujian Province. In their family records and village genealogy, they often emphasize their connection to Central China in expressing ethnic identity.

Despite the academic debates on the origin of Hakka (Xia 2011), the common Hakka people in Hongkeng Village did not consider Hakka identity as something that needed to be particularly emphasized in their daily lives before the recent development of tourism. For most of the time, Hakka identity was more a subconscious issue; it was neither clearly defined nor publicly celebrated within most of the communities. "I didn't have knowledge about Hakka in the past. I started to learn about Hakka after joining the tourism business. I learn from books. The tour company also provides us with training courses. I also even learn from tourists," Liping, a Hakka resident in Hongkeng Village, confessed, noting that it was hard for her to introduce Hakka culture for her job as a local tour guide. The promotion of Hakka ethnicity for cultural recognition and touristic development has brought the local people into more frequent and direct contact with the issue of identity and ethnicity. The touristic practices not only evoke local people's consciousness of and emotional and cultural attachment to Hakka identity but also invoke their strategic exploitation and construction of that identity.

In August 2008, one month after Fujian Tulou was designated as a UNESCO World Heritage, Yongding started the grand project of constructing the Hakka Exposition Park (Kejia Bolan Yuan, 客家博览园) in an eastern suburb of Fengcheng Town. The project, with an estimated investment of more than RMB 1 billion (about USD 170 million), was finished in 2012. The park consists of the Hakka Folklore Performance Center, Museum of Hakka Culture, Memorial of Hakka Migration from Central China, Tulou Art Center, Hakka Culture Research Center, Tai Chi Center, Harmony Wall and another nine subprojects to present Hakka and tulou culture. Many of the structures borrow tulou architectural elements.

Due to the competition for resources of Hakka ethnicity, Yongding County was careful in choosing the name for the park. At the beginning, the former secretary of the Yongding Municipal Party Committee proposed the title of Hakka Temple (Kejia Shengdian, 客家圣殿). Later this name was thought to be inappropriate as Yongding is not the oldest settlement of Hakka ancestors like places farther north in Fujian. The government also abandoned the more secular name of Hakka Cultural Center (Kejia Wenhua Zhongxin, 客家文化中心) because there are already several Hakka cultural centers in neighboring regions such as Meizhou and Ganzhou. In addition, another Hakka cultural center was under construction in the neighboring county Shanghang. Finally, the decision was made to name the project the Hakka Exposition Park for the purpose of distinction. From this detail, we see how the local government elicited the distinctiveness of Hakka ethnicity in Yongding while carefully avoiding controversies.

The construction of the park was an exercise in creating symbols and meanings that reflected Hakka heritage. During the final stage of the construction in August 2011, the Yongding County government sent people to collect soil from Henan Province in Central China, which is regarded as the homeland of the Yellow Emperor (*Huangdi*, 黄帝), the legendary ancestor and initiator of Chinese Han civilization. They also collected soil from several Hakka-concentrated communities in mainland China, Hong Kong, Macau, and Taiwan as well as cities across five continents with a large number of Hakka immigrants, such as London, San Francisco, and Vancouver. The collected soil was used for constructing the Harmony Wall to symbolize the Hakka people's common roots and their harmonious coexistence with and in other cultures. The symbolic structure was intended to help build connections between Hakka in Yongding and Hakka in other parts of the world while tracing their roots to the Han Chinese in Central China. As in many architectural forms, the dimensions of space and time are manipulated to create connections and meanings.

Another structure in the park, the Passage of Hakka Family Name Tablets, creates connections through genealogy. Despite a complex and long history of migration and immigration, the Hakka have been traditionally oriented around the notions of lineage and kinship (Constable 1996; Watson 2004; Peng 2006). While connecting Hakka people to their ancestors who migrated to the south from Central China, family names displayed on the tablets also provide a factor and a symbol that ties together the current Hakka population around the world as well, as expressed in the epilogue of the tablets: "Even though a tree is tall, you will find its roots; even though a river is twisting, you will find its origin.

Family name is a label of humanity and a carrier of civilization. Every generation of Hakka respects its ancestors in a reminiscent way. The tablets help us find our roots, connect those related, encourage us to follow our ancestors' spirit and educate the younger generations." In a heritage and tourist space, the lineage focus of the culture is also purposefully linked to the architectural form of tulou. The preface for the tablets' installation notes:

> Family name is a symbol of a clan. The development and formation
> of family names in the long history have become a significant part of
> Chinese culture. . . . Hakka is an outstanding subgroup of Han. Western
> Fujian is one of the major Hakka settlements. And the Yongding Hakka
> attracts attention from the world for their creation of the architectural
> wonder tulou. Nowadays, the Hakka Yongding are politically stable,
> economically prosperous, and culturally thriving. Thus, the tablets of
> Hakka family names are presented to express the fundamental cultural
> influences on tulou, to promote family name culture, and to carry forward
> Hakka spirit.

On the one hand, Hakka people identify themselves with Han. On the other hand, they proudly distinguish themselves as an "outstanding subgroup" that shares the glorious Chinese tradition while also possessing distinctive cultural features. Particularly, the Yongding Hakka people regard themselves as standing out with their construction of tulou, which is a manifestation of broader Hakka spirit. As asserted by local scholars, tulou were originally built to combat the harsh natural environment and to protect Hakka as new settlers against attacks from other groups. In this discourse, then, the Hakka lineage system is the social foundation for the construction of the communal architecture.

CONCLUSION

This chapter has offered the historical, social, and special settings of the Fujian Tulou World Heritage Site examined in the book. The chapter provides a synthesized presentation of the local time, space, knowledge, and social relationships that weave together to form the geographical and sociocultural environment and context that condition the formation of tulou as vernacular architecture and the use of this architectural form in pursuit of current cultural and socioeconomic goals.

The local society is experiencing a change from being a remote and relatively isolated area to a more accessible one due to economic and tourism development. The local government has great ambition for urbanization and modernization as shown in its long-term strategic plan. The area's development and increasing connection with the outside world changes local consciousness and perception of local community and culture. As some communities have started to transform from pastoral to nonagriculture-dependent societies, new opportunities and lifestyles have emerged within these communities. In this context, local government and the communities are more actively engaging in the presentation, representation, and, in some cases, invention of the historic context of the area for outsiders. These displays and constructions are intertwined with local narratives, shared memory and history, and community members' sense of place, all of which feature Yongding County and Hongkeng Village as cultural resources because of their distinctive Hakka and tulou elements.

NOTES

1. In the history of China, the central governments often referred to people who lived outside the territory of Central China, especially the non-Han groups, as barbarous or savage (*manyi*), perceiving them as less civilized. The land where the Hakka people live now was referred to as the land for the southern barbarians (*nanman*) in history.

2. The source for this story is from a digital version of *Yongding Gazetteer* provided by the director of Yongding Gazetteer Committee on July 8, 2012. As a tradition in Chinese history, gazetteers chronicling local communities were regularly written and updated by local intellectuals and many are currently well preserved in archives. This genre of local documentation provides a rich source for the historical understanding of Yongding and Hakka culture.

3. These statistics and information are from *Yongding Gazetteer* (1988–2000) provided by the Yongding Gazetteer Committee.

4. Yongding County Bureau of Urban and Rural Construction and Planning is a governmental agency that is in charge of significant aspects of the county development such as setting goals for the county development, regulating and managing the real estate market, making industry policies, and supervising industry management. It also plays a significant role in the management and development of local tourist destinations.

5. The source for this information is public information available in the "Strategic Plan for the Cities of Yongding County in Fujian Province (2011–2030)," posted by Yongding County Bureau of Urban and Rural Construction and Planning in 2012.

6. The source for this information is public information posted by Yongding County Bureau of Land and Resource in June 2012.

7. In the Chinese administrative division system, township is a unit of local government within counties. It usually consists of a town, which is the administrative center and a more populated area, and villages.

8. Hsieh T'ing-yu (1929) summarized five theories that attempt to explain the origin and history of the Hakka people: "(1) The Hakkas are descendants of the Mongol garrison soldiers. (2) The Hakkas are aborigines from Fujian who had been assimilated by the Chinese. (3) The Hakkas are descendants of the half million soldiers sent by Qin Shihuang, many of whom intermarried with women from ethnic groups in southern China. (4) The Hakkas are descended from the remnants of the Kingdom of Yue that was destroyed by [the state of] Chu in 333 BCE. (5) The Hakkas are descendants of Chinese coming from the Northern section of China following the Jin and Tang dynasties." For details, see Hsieh, 1929.

TWO

—␖—

TULOU AS VERNACULAR
ARCHITECTURE

TRADITIONAL ARCHITECTURAL DESIGNS and techniques have been greatly trumpeted in local promotion of tulou as well as in heritage and touristic presentation and narratives. This chapter focuses on the knowledge and practices surrounding the materiality of the tulou, examining its construction process, architectural form and style, and function and meaning. Tulou was designed to serve local people's needs based on traditional architectural technologies and locally available resources. Local natural and social settings played a significant role in the emergence and development of tulou as a local architectural form. Connecting to the Hakka history described in the last chapter, local people and tour guides often point out the defensive features of tulou, such as the thick walls without windows on the first and second floors and the thick ironclad gates. The defensive structure and function of tulou architecture are said to have been designed for the harsh social and environmental circumstances described in chapter 1. At the same time, they also tell tourists that the function of defense is no longer necessary in the present day.

In addition to the skills and traditional knowledge of their makers, the physical design and spatial use of tulou are closely related to the natural, cultural, and social conditions of the local community. As Michael Ann Williams points out, architecture is not only "a geometrical object" but also "an ordering of empty space"; we need to study human and social culture in order to understand the object as well as the ordered space (1991, 37). Thus, this chapter also examines how the local environment, natural resources, forms of social organization, and local values and beliefs relate to the design and construction of tulou as massive communal dwelling structures. Furthermore, this chapter illustrates how local people negotiate living environments, space, and social relationships through the physical form of tulou.

Large tulou were no longer built in Yongding after the 1980s as single-family homes became more popular and modern architectural materials and techniques were introduced to the area. My account of the building of large tulou and the associated ritual of tulou construction are based on observations during my ethnographic research, oral accounts from local informants, and materials provided by local scholars of tulou and Hakka culture.

TERMINOLOGY

The terms and definitions of tulou are an unsettled issue. Tulou, which literally means "earthen building," is also translated into English as "rammed-earth building" or "tulou houses." In accordance with scholarly and UNESCO statements, most people use the term *tulou* to refer to the southwest Fujian form of large houses in which the major structural element is rammed earth. However, even for the natives of the area, the definition of tulou remains ambiguous. Building houses with the technique of ramming layers of earth was common in western Fujian Province (Qian and Yin 2014; Xie 2004). Besides the expansive, multistory earthen buildings like those found in Hongkeng Village, which are intended for large extended families, there are also small rammed-earth buildings of one or two stories for single families. Some people consider all houses made of rammed earth as tulou, while others insist that only the large buildings of at least three stories can be categorized as tulou. Ziwu's sisiter-in-law in Yucheng Building, Chunhong, told me that they used to only use tulou to refer to those big circular structures. Some people whom I interviewed in Yongding County declared that all tulou were built by Hakka people, while some had a different opinion, pointing out that some tulou outside of the Hakka district were not built by Hakka people.

Responding to the various definitions and understandings of the term *tulou*, local government and cultural agencies intervened and classified tulou based on their sizes. This classification, as part of the cultural institutionalization of tulou, and similar to the effect of Intangible Cultural Heritage identification, recognizes and highlights the tulou buildings that are regarded by agencies with various authority as worth paying attention to. The Yongding County government categorizes tulou into "ordinary tulou" (*putong tulou*, 普通土楼) and "featured tulou" (*tese tulou*, 特色土楼). The ordinary tulou refer to the smaller homes with only one or two stories, while featured tulou refers to the large buildings that are at least three stories tall and that provide communal dwelling space for multiple households. In this case, all buildings made of rammed earth regardless of size are called tulou, but there is the issue of representativeness by selection. According to Yongding County official statistics, there are approximately 22,000 ordinary tulou in the county and approximately 1,000 featured tulou. Among the featured tulou, 362 are in a circular shape.[1] Compared to the ordinary tulou, the featured tulou attract much more attention and are more likely to be targeted as political, cultural, and economic assets. They have been carefully documented in the county archive and marked on the county maps. The featured tulou are generally regarded as the symbol of Hakka ethnicity, although the ordinary tulou are much more common across the Hakka region in western Fujian Province and far outnumber the featured tulou. The efforts to define and classify tulou are part of the institutionalization and standardization process that is necessary for tulou to be selected as a recognized form of cultural expression.

In addition to the size and representativeness of the architecture, ethnic association is another site of contest in local discussions. The topic of whether tulou are originally built only by the Hakka people is frequently brought up by local residents and by various agencies. Scholars who are regarded as the authority on the matter play an important role in the debate. People often make reference to them or their works when talking about the origin of tulou. But it is obvious that those who participate in the debate on the issue are unable to reach a consensus. Hu Daxin (2006), a local scholar and the director of Yongding Museum, claims that tulou is any building of two stories or more with earthen walls and wooden beam columns as the major structure. He further asserts that tulou as a unique architectural form first emerged and developed in Hakka society and was later transmitted to adjacent regions (2–3). However, Huang Hanmin (2009), the director of the Fujian Architecture Design Institute who has studied tulou since the 1980s, argues that it is inappropriate to claim that the origin of tulou is in the Hakka region, as tulou can also be found in other regions. He defines tulou as large, multistory buildings in southwest Fujian mountainous region that are intended to provide for a large community dwelling and are conducive to defense. In his understanding, they are built with weight-bearing rammed-earth walls and a wood frame structure (112).

While the origins and definition of tulou are still unsettled, a new term *Fujian Tulou* was created to specifically refer to the residential earthen buildings in western Fujian Province that are designated by UNESCO as World Heritage Sites. In terms of architectural features, one noticeable difference between Hakka tulou and Minnan tulou in Fujian Province is that Hakka tulou usually have more shared public space and put less emphasis on privacy while many

Minnan tulou have more enclosed or semi-enclosed space within each unit or several units so as to provide more private space for the residents living in the unit(s). Despite the architectural difference and the contested discourse, under the provincial and national leadership and with a common goal, the local government of the Hakka side and the local government of the Minnan side reconciled and worked with joint efforts and resources for the nomination of tulou as UNESCO World Heritage Sites. During the nomination process, the term Fujian Tulou was applied to resolve the dispute. The term eliminates the group association of tulou to a specific ethnic group and governments of the Hakka region and the Minnan region. The regional vernacular architectural term *tulou* is defined by the larger geographical term *Fujian* to expand the political-geographical boundary and put the Hakka tulou and Minnan tulou in the same category with the satisfaction of both the Hakka side and the Minnan side. Since 2008, Fujian Tulou has become a legitimated term used by UNESCO and governments to refer to the designated featured tulou and tulou clusters (map 2.1). These tulou, according to UNESCO assessment, are "the most representative and best preserved." The UNESCO web page explains that "Fujian Tulou is a property of forty-six buildings constructed between the 15th and 20th centuries over 120 km in southwest of Fujian Province, inland from the Taiwan Strait."[2] Consequently, many people tend to think of tulou as those universally recognized massive structures, although Fujian Tulou as a term is not commonly used in local people's daily lives. Scholarly studies and official definitions of tulou became prevalent in the process of tulou World Heritage nomination and heritage-related tourism development. This in turn has heavily influenced local people's current understandings and perceptions of tulou.

In addition to tulou, there are other forms of Hakka residences such as *weiwu* (enclosed house, 围屋) found in adjacent southern Jiangxi Province and *weilongwu* (enclosed dragon house, 围龙屋) in adjacent northeastern Guangdong Province (Jiang 2004; Poon 2009; Yu 1997). Like tulou, weiwu and weilongwu are also massive enclosed structures with defensibility (usually one entrance to guard and thick, strong walls) and communality (space and facilities shared by many households). Weiwu (also called Hakka weiwu) are often in square or rectangular shapes with watchtowers at the four corners. Some weiwu are made of rammed earth, and some are made of bricks or stones. Weilongwu usually refers to the semicircular structure with a semicircle pond in the front (fig. 2.1). It can consist of one to three semicircle structures, and its design and arrangement of space is significantly different from those of tulou. However, some Hakka people in Guangdong Province include tulou in the category of weilongwu. Alternatively, some people also categorize weilongwu as tulou.

Map 2.1 Location of tulou and tulou clusters inscribed as UNESCO World Cultural Heritage and referred to as Fujian Tulou. (Map created by author based on material provided by the Archive of Yongding Office of Cultural Heritage)

Inspired and encouraged by the successful nomination of tulou, governments in Guangdong have started to work on the nomination of weilongwu as UNESCO World Heritage Sites ("Meizhou City Initiated Hakka Weilongwu Heritage Nomination" 2016).

Since the focus of my work is to examine the UNESCO-inspired heritage and tourism phenomenon in Yongding as a specific community, the UNESCO-defined Fujian Tulou is an important reference to the use of tulou in my work. As I also examine the local community's daily activities as a whole, and my fieldwork includes both the designated tulou and the nondesignated tulou, I do

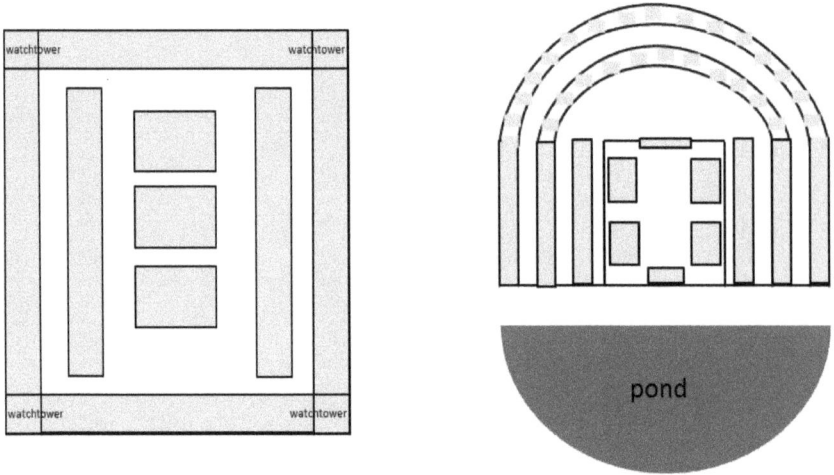

Figure 2.1 Illustration of the structure of weiwu and weilongwu.
(Illustration by author)

not exclude those tulou that are not on the UNESCO list. However, because the phenomenon such as heritagization, tourism development, and cultural presentation I study here has only been caused by the featured tulou, my main focus in the study is the massive structures that attract the most national and international attention rather than the small single-family houses. Therefore, *tulou* in this study refers to the large-scale distinctive vernacular architecture structures built of earth and wood and used for communal dwelling in the region of southwestern Fujian Province.

THE CONSTRUCTION OF TULOU

The construction of tulou was a grand project that required cooperation and strategic planning.[3] It was also a long process that could take several years or even more than a decade to complete. Thus, the builders needed to carefully calculate expenses and consider the design of the structure. When considering where and how to build a residence, the dwellers must think of the most convenient and safe place to begin construction and the best way to make use of readily available natural resources from the environment around them. Generally speaking, it is easier to build on land that is already flat; thus, in a mountainous area like Yongding where flat land is limited, those who competed for space needed to figure out a way to coexist. Hence, as demonstrated in this chapter, the construction process reflects the native social relationships, morals, and values relating to strategic negotiation as well as cooperation and compromise. Furthermore, once the Hakka people secured space in the natural world to build massive residences that were expected to be durable enough to house their families and descendants, rituals had to be performed properly, and deities had to be invited to protect the construction process. Therefore, construction involved adjustment to the natural surroundings, cooperation and negotiation between the housebuilder and his neighbors, and mediation between the human world and the supernatural world.

As previously stated, flat land was one of the most precious commodities in the area's hilly terrain. It was oftentimes not easy to get a large piece of land in the valleys for constructing tulou, especially when the choice of construction location was confined by local folk beliefs of *feng shui* (wind water, 风水) principles, the geomancy popular in Chinese vernacular society that determines an auspicious location for a structure with reference to local features so as to harmonize human existence with the surrounding environment and ensure the well-being of the residents.[4] The most favorable location for a house was with the azure dragon on the left, the white tiger on the right, the red bird in the front, and the black tortoise in the back; more specifically, the ideal geographical configuration was that the house faced an open space with a river running in the front and was surrounded by hills on the other three sides. The dragon, tiger, and tortoise are all metaphorical terms referring to hills. I often heard people proudly tell me their house was located in such a favorable and auspicious place. In turn, people believe that the location will ensure or strengthen the safety, prosperity, and honor of their family.

It often required strategic planning and wisdom to obtain a piece of favorable land or just enough space for the house construction. Many local narratives tell stories of how Hakka ancestors strategically got the land they wanted for tulou construction with patience, wisdom, and skills in interpersonal relations. Jiang Linghong, a resident of the "king of tulou" Chengqi Building, told the story about the process of land acquisition for the construction of Chengqi Building (fig. 2.2).

> Our ancestor Jiang Jicheng didn't get a formal education because his family could not afford it. But he was very smart. He raised ducks to make a living. He often took his ducks to the fields near where Wuyun Building [a tulou that is situated very near the current Chengqi Building] is located. He thought it was an auspicious place and wanted to settle here. Thus, he bought a small piece of land and built a small hut there.

Figure 2.2 Chengqi Building, which is nicknamed "the king of tulou" of Gaobei Tulou cluster at Gaobei Village, Gaotou Township of Yongding County, Fujian Province. (Photo by author)

Some people were suspicious about his purpose of building the hut. He explained that he raised ducks in the area and needed a place to rest.

The original owner of Wuyun Building was rich. However, when he passed away, his sons were lazy parasites. They often lost money in gambling. Jiang Jicheng treated them nicely and often invited them to his hut for tea and food. He also lent them money to pay the gambling debt. As time passed, it became difficult for those young people to pay the money back. Finally, they offered to sell Wuyun Building to our ancestor. With the money he saved and borrowed from his brothers, our ancestor bought the building.

However, some former residents of Wuyun Building who had already moved out said the building was built by their common ancestors and they would not like to sell it. They tried to tear down the newly built wall while Jiang Jicheng was rebuilding the house. On one hand, Jiang Jicheng asked his wife to treat those people with tea and snacks; on the other hand, he urged the workers to speed up the construction. Those people who came to tear down the wall felt too embarrassed to do so as they were treated so nicely. And the construction was completed in several days. According to Hakka custom, once the construction finishes, it will be regarded as disrespectful to intervene.

After Jiang Jicheng's family moved to Wuyun Building, his four sons brought him twenty grandsons. Two of his sons became local officials and eleven of his grandsons became local scholars. Our ancestor's family expanded and became more prosperous.

By the fourth generation, Jiang Jicheng had more than seventy descendants. Wuyun Building is not big enough for such a large family. Jiang Jicheng made money through planting tobacco and buying land in other villages and leasing it to local peasants. With enough money saved, he wanted to build a new house. Thus, he bought most of the land where Chengqi Building and Shizhe Building are now located.

To construct Chengqi Building, Jiang Jicheng acquired most of the land, but the owner of a piece of land in the middle refused to sell his land. So, our ancestors had to build the first floor of the outer wall first. Then Jiang Jicheng went to negotiate with the landowner: "Would you please sell us your land? We already built the first floor. It will be difficult for you to irrigate your land." The landowner said he could bring the water through bamboo pipes set above the wall. Our ancestors had to continue increasing the height of the outer wall. As a result, the bamboo pipes for irrigation had to be lifted as the wall became higher and higher. Finally, the wall became too high for the landowner to bring water to irrigate his crops. At that moment, Jiang Jicheng took lavish gifts to the landowners' place and nicely asked him whether he would like to exchange or sell the land. As our ancestor was a good man and had been treating the landowner very nicely, the land owner agreed to sell the land to him. Our ancestors finally got the last piece of land for constructing Chengqi Building; it took more than a decade to finish the whole construction.[5]

Jiang Jicheng, the ancestor of the residents in Chengqi Building, was born in 1635. He built three tulou and four home schools for his family and descendants. In the hearts of Jiang's descendants, this ordinary farmer is a legendary hero. Now in the twenty-seventh generation of his descendants, all the residents are familiar with Jiang Jicheng's stories and feel proud of their ancestor. In addition to the character of being diligent, persistent, and cautious in managing and spending money so as to save enough to build tulou, he is greatly admired by his descendants for his skills of negotiation, patience, and wisdom. The narrative also clearly demonstrates the cautious interactions he pursued with other residents and landowners in the same community. Similar narratives can be found in Hongkeng Village and such narratives gain more space for presentation and circulation in the circumstance of touristic practice.

As previously stated, it was difficult to obtain a piece of land in the valleys and local beliefs and taboos constrained people's behavior in the process of land acquisition. Thus, the design and shape of tulou were often conditioned by both objective and intersubjective circumstances. In heritage and tourism contexts, the most promoted tulou are the ones that are round in shape. Those with square or rectangular shapes are regarded as less unique. Tulou that are irregular in shape, such as oval-shaped, semicircular, U-shaped, or octagonal, are selected to represent variations of tulou. A story locally transmitted about Dongcheng Building in Nanjiang Village illustrates how local beliefs, taboos, and social relationships affected the design and shape of tulou.

> Dongcheng Building is known for its octagonal shape with eight corners. It is a three-story building of approximately 4,200 square meters. When it was built in the mid-nineteenth century, the building was originally designed to be square in shape. However, one of the four corners pointed at a tomb, which was regarded by its owner as "a tiger dashing down a mountain." The corner of the building looked like an arrow aimed at the head of the tiger, which was believed to be bad for the tomb owners. Thus, the tomb owners obstructed the construction of Dongcheng Building.
>
> When the tomb owners and the building owners got into fierce conflict, somebody suggested they invite the famous retired scholar of the imperial academy, Wu Yifu, as a mediator to resolve the problem. Wu Yifu carefully investigated the case. As a native [of the area], he understood that it was one of the greatest taboos that a tomb be pointed at by something like an arrow such as the corner of a house. And he also acknowledged that the building owners had no other choice but to build their house at the current location. He talked with both sides respectively. Then he invited both sides for a meeting. After a speech about the importance of maintaining harmonious relationships among neighbors, he suggested the building owners cut the corner facing the tomb. He also suggested they cut the other three corners to achieve a balanced design. The building owners accepted his suggestion and built the octagonal Dongcheng Building.[6]

As suggested by this house legend, problems could arise even before the actual building of tulou. In the process of negotiation and solving the problems, the previous narrative emphasizes the significance of maintaining harmony among community members and how it took deep understanding of local beliefs, wisdom, and compromise to achieve this goal.

The actual construction process of tulou involved not only human interaction and relationship but also continual communication with and mediation of the material, social, and supernatural worlds. In local conceptions, people believe that *tian* (universe, which often implies the supernatural world of gods and deities and some other realm or space that is out of the reach of human beings), *di* (earth, which usually refers to the natural environment and the material realm), and *ren* (man, which means the individual and the social society) should be appropriately positioned so as to achieve harmonious coexistence. In local narratives, especially the current touristic narratives, the construction of tulou is oftentimes regarded as a process of communication and negotiation between tian, di, and ren.

Hakka people in western Fujian Province customarily used organic materials such as earth, wood, and bamboo for building construction as they were locally available and plentiful. Rocks for building foundations and flooring were collected from nearby rivers and hills. All timber structures of a tulou were made of the Chinese fir trees that grow commonly in the area. Fir trees were cut down and dried with the bark removed long before the construction begins. Moso bamboo, like fir trees, was one of the major plant types in the region. Bamboo splints were cut and dried for making "wall bones," acting like steel rebar in concrete, to reinforce the earthen wall. The earth used to ram the wall was a mixture of clay and sand. As the buildings were exposed to flooding, lightning, fire, and earthquakes, which are devastating to them, some households added a certain amount of lime to the mix of clay and sand to enhance the hardness and lower the permeability of walls, especially in areas subjected to flooding.

After preparing the construction materials, people tended to start building tulou in the second half of the lunar calendar year when the rainy season was over and the busiest farming time had passed. The actual construction process began with an important ritual. Before the construction started, a feng shui master was invited to select an auspicious occasion for the commencement ceremony. In the ceremony, the feng shui master set Yanggong, which is a deity of feng shui usually manifested by a bamboo or wood stick, at the two ends of the house axis line and led the house owners and the housebuilders to burn incense and to offer sacrifices of pork, beef, and lamb to Yanggong. Only after the ritual could people start to dig a pit for the house foundation to the sound of bursting firecrackers.

Yanggong was an actual historic figure named Yang Junsong who was born in 834. He was a high official in charge of the design of palace architectures, celestial observatories, and royal ceremonies in the palace. After the capital was taken over by rebels, he chose to live a reclusive life in the mountains and focused on the study of feng shui. He helped people with their house constructions and trained apprentices with the knowledge of feng shui in southern Jiangxi Province, the Hakka settlement area next to Fujian Province and Guangdong Province. People believe that feng shui gradually became popular in Jiangxi, Fujian, and Guangdong Provinces and then spread to many other parts of the country because of Yangong. Thus, Yanggong is regarded as the founder of feng shui who brought royal and privileged geomancy knowledge to vernacular society. The popularized feng shui deeply influences Hakka people's beliefs and practices in daily life. Yanggong is widely worshipped by Hakka people (Chen 2002; Wen 2006). The actual Yanggong used in the commencement ceremony is not a figurine of the real person but a piece of bamboo or a wood stick that is about one meter long. The stick is marked with the name Yanggong. Other legendary figures, including Lu Ban and He Ye, the founders of carpentry, are symbolically marked on the stick. Zhang Jiangu and Li Dingdu are regarded in some parts of China (including the Hakka region) as the attesting witnesses, sureties, or mediators of land lease agreements between the human world and the god and ghost worlds, and they are involved through two sticks that mark the axis line of the house to be constructed. As inscribed on the stick, the figures are invited to ward off ill luck and evil, to ensure safety during the house construction process and to bless the whole family. Upon the completion of the construction, some families enshrine Yanggong in the central hall of tulou to continue to protect and bless the family (fig. 2.3).

The next step after marking the house axis with Yanggong is to lay the five elements stones (*wuxingshi*, 五行石) in the foundation located at one end of the axis line (table 2.1). The five elements stones are stones of different colors and

Table 2.1. Colors and shapes of rocks that represent each element in the universe.

Shape	round	rectangle	pentagon	triangle	square
Color	white	green	black	red	yellow
Element represented	metal	wood	water	fire	earth

Figure 2.3 Residents offering incense and food to Yanggong made of bamboo stick enshrined in the central hall of Yucheng Building in Hongkeng Village during Mid-Autumn Festival. (Photo by author)

shapes that represent the five elements in Chinese geomancy—wood, earth, water, fire, and metal. People place the stones in the foundation with the hope for a prosperous and thriving family. The custom is common in the construction of different forms of Hakka vernacular buildings, although the actual practice varies in different areas. For instance, some people bury the stones in the foundation underground while in some Hakka areas in Taiwan, people place the stones on the surface of the foundation above the ground (Liu 2005).

After the ritual of placing the symbolic stones, the builders started to lay the foundation with rocks. The height and width of the foundation depended on the geographical location and soil properties of the land on which the tulou was constructed. Solid ground provided a stronger support to the foundation on which the wall was built. So, a shallow foundation was often sufficient to sustain the weight of thick walls. However, a deeper and thicker foundation was required when constructing tulou on a relative soft ground. The oldest tulou on the northern side of Hongkeng Village is Dongsheng Building, which was built on a piece of silty, rather than solid, land. The constructors put pine logs in the loose soil and filled the land with gravel to form a solid foundation for the building. After about 350 years, the building wall has slightly inclined, making the building a leaning tulou, but that doesn't seem to pose concern or risk to the five residents who still live there. For many tulou, the bottom part of the wall was built with rocks rather than rammed earth to prevent potential damage brought by flooding. In flood zones, the rock bottom is usually three to six feet above the ground so as to prevent the earthen wall from being waterlogged.

The foundation was let to sit for at least half a month after completion to ensure that it was fully hardened. Then the house owner asked a feng shui master to choose an auspicious moment to set the board molding frame (*qiangfang*, 墙枋) on the foundation or the rock base before the construction of rammed-earth wall. The board molding frame consists of two pieces of thick wooden boards secured by cross-staves on one end and the width between the wooden boards can be adjusted (fig. 2.4). When the board molding frame was first placed on the foundation, the family members offered sacrifices in front of it, including pig, cattle, and lamb, and prayed for safe and trouble-free construction of the earthen wall.

The builders started ramming the earthen wall either from where the five elements stones were placed or from the auspicious location indicated by the feng shui master. Earth mixed with other materials such as sand was poured into the board molding frame. Builders standing in the board molding frame rammed the earth repeatedly with a wooden pestle of approximately six feet long. Strips of wood or bamboo splints were laid in each layer of the rammed-earth wall for reinforcement. As noted previously, the wood or bamboo splints are called "wall bones." This is a technique that is boasted of proudly by local people and scholars. A cross-bonding technique was applied in the positioning of the qiangfang for strengthening connected layers of the rammed earth. The builders usually rammed no more than three layers each day so that there would be enough time for the wall to dry. The people who rammed the wall also use wooden pounders to flatten the two sides of wall by pounding the uneven places. The thickness of the wall decreased gradually as it goes higher, and the wall was slightly inclined inward so as to lower the center of gravity in the wall as a whole for better stabilization.

When the builders completed ramming the wall and started to set up posts and beams, the house owners celebrated by putting couplets of auspicious words written on red paper on the posts and beams. The roof beam right above the central hall, which was regarded as the most important beam, was wrapped

Figure 2.4 Tools for making the rammed-earth wall of tulou. On the right end of the board molding frame (qiangfang) are the "wall bones" of bamboo and wood, and on the left end (secured by cross-staves) are a wooden pestle and tamping tool. (Photo by author)

with red cloth and hung with crop bags and books. The house owners also prepared rice cakes to treat the builders and offered sacrifices such as pork, lamb, chicken, duck, fish, tofu, and rice cakes in front of the beam. The feng shui master, carpenters, and builders who did the construction received red envelopes with monetary gifts when they said words of blessing to congratulate the house owners.

Tiling was the last step for constructing the main structure of tulou. It started from the bottom of the eaves. Each upper tile covered seven-tenths of the lower tile. Usually two-meter-long eaves stood out to protect the earthen wall from rain. The major construction of tulou was completed when every part of the roof was tiled. At that moment, the house owners showed their gratitude to the deity Yanggong with sacrifices for a safe and successful construction process. As mentioned, some people enshrined Yanggong in the central hall.

Others might just burn the deity's wood or bamboo form so as to send it back to the heavenly realm.

The gate of the tulou was set up when the construction of the major structure was completed. The gate was regarded as of great significance to the well-being of a family. Thus, it needed to be carefully positioned. A feng shui master was invited to calibrate the position of the gate with a compass. And the gate had to be set up at the chosen auspicious time, with sacrifices offered to the door god.[7] The construction of the kitchen stove as another important part of a house, one that was also ritualistic, with many symbolic acts and items. Once more, the feng shui master was invited to choose an auspicious time and place to build the stove. The seeds of crops and coins were buried at the four corners of the chosen site. Sacrifices were offered to the kitchen god.[8]

Once the kitchen stove was built, the residents could prepare to move into their new house. Before they moved in, however, the feng shui master performed an exorcism in the building to ward off evil and ill luck. After the house was purified, the residents lined up according to their age, with the oldest person in the front. Each held a household item such as a broom or a pot, and they happily moved into the new home.

As previously described, the construction of tulou was not only an act of physical space arrangement but also a process of communication and mediation between the natural world, human society, and the supernatural world. The housebuilders, as craftsmen, transformed the natural material to create a dwelling space for the residents. The feng shui master played the major role of mediating between time, space, nature, humans, and gods. Feng shui was a sophisticated system that people relied on for expressing hopes and wishes as well as seeking physiological and spiritual support during the construction of tulou as a massive, high-stakes project. People believed that when they built the house according to the auspicious time and location chosen by the feng shui master, the structure would become a blessed space, one that would bring good luck to the family or ensure their well-being. Symbols created and rituals performed during different stages of construction were also for the same purpose of embodying human expectation and spiritual elements in the physical structure. The construction process strengthened the residents' emotional and spiritual connections with the newly created space.[9] Even though no new tulou are being built in the local region, the skill of tulou construction was inscribed on the First National Intangible Cultural Heritage List in 2006, which recognized some master tulou builders as representative inheritance persons of tulou construction skill.

STRUCTURE AND FUNCTION OF TULOU

Yucheng Building, the tulou that I stayed in during my visit at Hongkeng Village, is located in front of a hill by the Jinfeng River; it houses two extended families with a total of twelve people permanently living there. Privacy is mostly limited to the bedrooms on the upper floor(s). Every morning when I woke up in my bedroom on the fourth floor, I could hear sounds of cookware in the kitchen, residents chatting in the halls, children playing in the central yard, and sometimes my hosts' new baby crying in the dining room on the first floor; the sound permeability is due to the open inner space of the building. The person who got up first, usually one of the elders in the families, opened the big heavy wooden gate. The gate was kept open for the whole day. Tourists, and sometimes visiting villagers, came in and out of the building by crossing the horizontal piece of stone that forms the bottom of the doorway.

The design of tulou follows the principle of "closed outside and open inside," which means that it is enclosed by a thick wall that guards the residents against intrusion or disturbance from the outside world, whereas the inside is designed for all the residents to share common spaces rather than putting much emphasis on separating spaces for each household to protect personal privacy.

Viewed from the outside, most tulou are simple solid structures with plain outlines sketched by the beige walls and the black roofs, humbly and harmoniously merging in the surrounding natural environment. Conventionally, the first and second floors of tulou are solid with no windows or any other openings in the walls except the gate and (maybe) one or two side doors. Above the second story, there may be small holes in the walls that can be used as windows or, as it is said, "gun holes" during turbulent times. The gate is strong and defensive. It is made of thick, heavy wooden boards (sometimes reinforced by attaching iron armor plates to them) and is usually framed by large, heavy stone slabs. The gates are barred inside with horizontal wood posts inserted into holes in the stone slabs or the walls. Some gates are equipped with wooden tanks above it for storing water in case of fire.

The fortified and defensive features of tulou are believed to be related to its natural and social environment in the history of its emergence and development (Huang 2009). On one hand, southwestern Fujian used to be full of remote forests that harbored wild beasts that could harm the residents. On the other hand, these areas were plagued by armed bandits and occasionally rebels from the twelfth century to the early twentieth century, when official governance was too far removed to effectively protect the area. These architectural features are repeatedly emphasized by local scholars and also by tulou residents in the tourism context in their introduction of tulou. However, examining the history and architectural details of tulou in the social context during the Chinese Civil War from 1927 to 1949, Jing Zheng (2013) found that tulou's function has evolved. Tulou built in the early twentieth century were built as affordable cooperative and collective dwellings in rural areas of southwestern Fujian Province to solve the problem of population pressure at that time. In these, the defensive features found in tulou from earlier times were eliminated or simplified to reduce construction cost. The defensive function of tulou was diminished by modern weapons such as artillery. So, rather than using the architectural structure for self-defense, local people sought community security through political negotiations (Zheng 2013, 53). Regardless of whether tulou were originally constructed to be fortresses (which has never been fully confirmed), local tourism narratives continue to champion the defensive function of tulou.

In addition to defense and security, local experts and residents boast of other features of tulou material and design that contribute to comfortable living and environmental sustainability. The thick earthen walls are said to be windproof while also being able to regulate the humidity within the building by absorbing moisture during humid seasons and releasing moisture during dry seasons. There are also stories circulated in the communities about how the rammed earth and the wood and bamboo splints placed in the walls made tulou capable of resisting earthquakes in the past.

The interiors of tulou are much more decorative and sophisticated than the exteriors. The interior usually consists of private space and public space as well as secular space and sacred space. The private space consists of rooms distributed to each household, including kitchens, storage areas, and bedrooms. The public space is shared property among all the residents in the building, including the entrance hall, corridors, the central hall, the courtyard, stairs, and well(s). Most tulou are designed as symmetrical structures. The central hall is the most distinctive sacred space in the building and is always located on the axis line.

A tulou usually consists of three or four floors. The first floor of the outer circle is used for kitchens and dining rooms, the second floor is for food storage, and the upper floor(s) are for bedrooms. These rooms in the outer circle are distributed to each household living in the building as vertical set(s) from ground floor to top floor. Due to space limitations, rooms are quite small, with each room being approximately one hundred square feet.

Figure 2.5 A kitchen in Zhencheng Building with black ink characters written on red papers symbolizing the kitchen god on the wall above the stove. (Photo by author)

The kitchen is the only privately owned space on the first floor. It also functions as a dining room and living room in the case that a family gets only one vertical set of rooms. Conventionally there is no window on the outer wall of the kitchen so the upper half of the wall facing the corridor is made of slats with spaces in between for lighting. And in addition to the door, there is a half-height door made of slats. Usually, the half door is used to keep the poultry, which is raised in the building, away from the kitchen. Thus, the kitchen is a half open and half private space. The door will only be closed when the family is leaving home for an extended period of time.

The major facility in the kitchen is a large cooking stove that burns firewood and occupies a large portion of the small room. The chimney, made of pottery pipe, is buried in the thick earthen wall and the stack outlet located on the roof or the external wall of the second or third floor. Many families put a kitchen god on the wall of their kitchens (fig. 2.5). The kitchen god is usually presented with its title written on a piece of red paper. Unlike the deities Yanggong or

Guanyin enshrined in the central hall that is shared among all the residents and that protects all of the people who live in the building, the kitchen god is a family god that represents a family and protects an individual household.

Above the kitchen is the storage space for each family to store food and private household items. The storage is on the second floor for a good reason. First, because of the humid local climate, the first floor is too damp to store farm products such as rice, taros, and sweet potatoes. Second, it is less convenient to store food and household items on the third or fourth floor considering the narrow stairs in tulou. Third, the heat from the kitchen right under the storage room effectively reduces moisture, making it ideal to store farm products on the second floor.

Bedrooms above the storage are the most private space in tulou. Nobody is allowed to go into the bedrooms without the owners' permission. The door is not only closed but also locked in most cases. A small window is dug through the thick outer wall for lighting. There is also a bigger horizontally sliding wooden window on the wooden wall facing the corridor to let more light into the room and for better air circulation through both windows. The bedrooms are much drier and ventilated than the rooms on the first floors, which is be-lieved to be important for the residents' health while sleeping in a small space with a humid climate and without air conditioning.

The rooms on each floor are connected by shared corridors (fig. 2.6). Many corridors on the first floor are paved with a composite of lime-sand-earth or blue bricks. Corridors on the upper floors are made of wood boards. Some are also paved with a layer of thin bricks for the purpose of reducing noise and fireproofing. I saw only a few tulou with the corridors paved with bricks on the upper floor during my visit in Yongding County. There are usually two or four sets of wooden stairs in tulou. In rare cases in Yongding, stairs are at-tached to each vertical set of rooms, which provide a greater sense of privacy and independence.

The courtyard in the center of tulou is an empty, open space that allows air circulation and light into the building. There is usually a water well (or two) in the courtyard. It is not only a playground for the children but also a communal working place for the residents to wash clothes or clean other household items with water from the well. The entrance hall is another frequently occupied com-munal space with high flow of information exchange and communication. It is the residents' favorite gathering place, as light and breeze go right through the gate into the hallway. Thus, long wood boards or stone slabs are often put along the walls of the entrance halls to seat people. Residents like to bring their food in a bowl to eat here and chat with each other. Female residents often do their

Figure 2.6 Upstairs corridor in Zhencheng Building with a sign asking tourists not to proceed. (Photo by author)

housework in the hallway. It is a place for casual information exchange, rest, and play. In the early morning and after the sun sets, people also like to sit at the passage under the eaves of the front wall, with children playing in the front yard, which is usually paved with cobblestones.

The central hall is a sacred space for ceremonies and rituals. It is a highly symbolic and meaningful space for the residents. Located either in the center or at the backside of the axis line of the building, the central hall is regarded as the most important section in tulou, one that connects all of the residents in the building with spiritual attachment and clan/family identification. It is a shrine for ancestors and family gods such as Yanggong and Guanyin. Significant family or clan events such as ancestor worship ceremonies, festivals, meetings, weddings, and funerals all take place in the ceremonial hall. Announcements such as the birth of a new family member or the naming of a baby are also written on a piece of red paper and put on the wall in the central hall, while announcements of a more secular nature, such as notice of how common funds have been spent, are usually put on the wall of the front hall.

ZHENCHENG BUILDING: A REPRESENTATIVE
TULOU IN HONGKENG VILLAGE

I introduce one of the most celebrated tulou in Hongkeng Village—Zhencheng Building—here to illustrate the design and structure of tulou. Zhencheng Building is one of the largest and most popular tulou in Hongkeng and among all the UNESCO-designated tulou in Fujian (fig. 2.7). The construction of the double ring structure was started in 1912 and completed five years later. A fifty-two-foot-tall earthen wall of four stories forms the outer ring with a bottom that is four-feet thick. The inner ring is a two-story structure made of bricks and wood. The very center of the building is a shrine for ancestor worship (fig. 2.8). There are also two two-story brick-and-wood structures attached to the east and the west sides of the main structure that function as a workshop for processing tobacco and as a private school for home education.

The building is equally divided into eight sections separated by brick walls. The design is aligned with *bagua* (八卦) which literally means "eight symbols." Bagua is a fundamental and sophisticated cosmology/philosophy of Taoism that connects the living world with the natural and supernatural world. In physical form, bagua is marked on the compass that feng shui masters use to determine the position and the orientation of a house. The sign of bagua is eight trigrams formed with three broken or solid lines. The broken line represents *yin* (darkness 阴), which indicates the earth, female, darkness, and so on. And the solid one represents *yang* (brightness 阳), which connotes the sky, male, light, and similar. The physical manifestations of the concepts are often opposites, such as female and male, dark and light, water and fire, cold and hot, death and life, soft and hard, wet and dry. A bagua diagram shows the balance between yin and yang. Within this philosophy, yin and yang are opposite forces that need to be carefully balanced to maintain the best conditions or status in the real world. In tulou-related narratives and building introductions, harmony, as represented by this bagua design of tulou, is depicted as a fundamental philosophical concept that local people believe in and apply in dealing with their relationship with other human beings and with their living environments.

The outer circle of Zhencheng Building is the residential living space distributed to each household (figs. 2.9 and 2.10). There is a six-room space in each of the eight sections. The inner circle is a two-story brick-and-wood structure that serves as public space for meeting guests, and the second floor is also used as viewing stands for the audience when there is performance in the central hall. The outer circle and the inner circle consist of 208 rooms. Some small rooms made of bricks are bathrooms attached to the outer wall of the inner circle.

Figure 2.7 Exterior view of Zhencheng Building. (Photo by author)

Figure 2.8 Interior view of Zhencheng Building. (Photo by author)

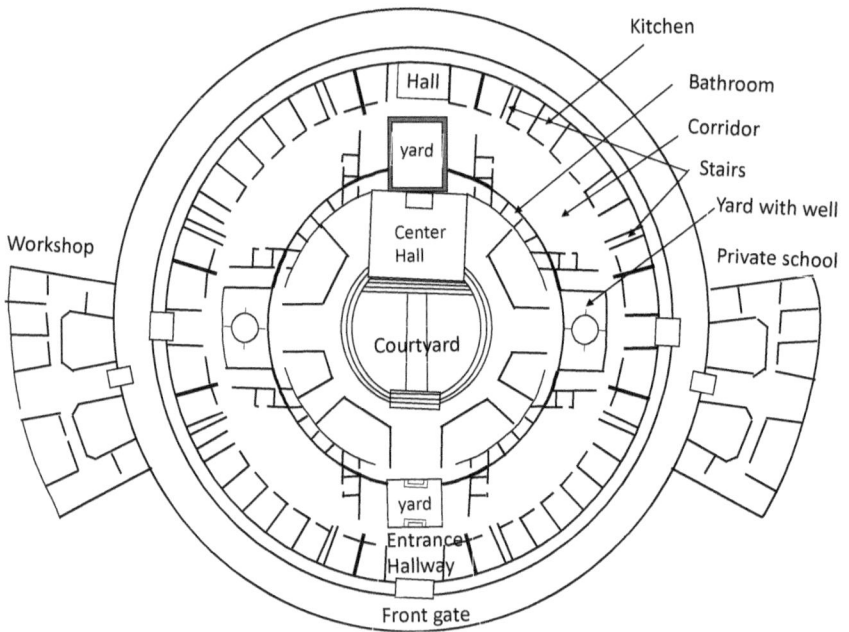

Figure 2.9 Sections of Zhencheng Building. (Modified figure based on materials provided by the Archive of Yongding Bureau of Cultural Relics with English labels added by author)

Figure 2.10 Floor plan of Zhencheng Building. (Modified figure based on materials provided by the Archive of Yongding Bureau of Cultural Relics with English labels added by author)

Along the symmetrical line are the front gate, the entrance hall, a small yard, a courtyard, a central hall, another small yard, and the back hall. The central hall is a shrine with the image of the residents' ancestors and words of moral doctrine on the walls. It is also a performance stage during special occasions such as festivals. Like many other tulou, a figurine of Guanyin is enshrined in the hall at the end of the symmetrical line. There is a yard on the left and the right side of the space between the inner circle and the outer circle. Two wells are located in the yards. One is called yin well and the other is called yang well. The whole building is neat and balanced despite its massive size and complex architectural structure. Zhencheng Building is now one of the most popular "scenic spots" that awe tourists from all over the world. Most of the sections that are commonly owned by the residents such as the passage on the first floor, the halls, and the yards are rented to the government to be open to tourists.

CONCLUSION

This chapter has focused on the materiality of tulou as a form of vernacular architecture for communal dwelling. Tulou and its construction are part of the local cultural and ecological systems. Skills, traditional knowledge, customs, beliefs, values, and social networks as well as the owner's personal traits are all integrated into the system. Construction materials are obtained locally with the lowest possible costs. Space is carefully negotiated to better serve the needs of individual families and communal life. Tulou were built to be shared by the extended family. Not only is the physical structure passed down from generation to generation, but so too are the stories associated with the construction of each tulou. The morality and wit of the residents' ancestor, as demonstrated in the acquisition of land and construction of tulou, are also passed down and circulated in the local community. Thus, tulou are both material and cultural legacies for the resident families.

Tulou as traditional dwelling is not only a tangible artifact but also a "cultural creation" (Glassie 2000, 18) integrating and embodying the intangible aspects of local beliefs, norms, and social relationships. Feng shui as the philosophical system that heavily impacts Hakka people's daily lives is the spiritual logic behind the planning and construction of tulou. Local people believe that the spatial-temporal positioning of an event or behavior has social consequences that affect their own lives and fortunes. The immediate situation of the planning of tulou requires not only will and wit but also adjustment and negotiation of human relationships. Henry Glassie (2000) sees architecture as "a kind of communication" that "shapes relations between people." I would add that, in the case of tulou construction, social organization and relations also shape architecture. Living in a tight place, people learn to wisely coexist with each other for a better life and living experience through the careful arrangement, ordering, and negotiation of space.

The materiality of tulou described here is vital in meeting the UNESCO World Heritage nomination criteria of being exceptional in terms of "size, building traditions, and functions" as well as "representative of a culture (or cultures), or human interaction with the environment" and reflecting "society's response to various stages in economic and social history within the wider region" (UNESCO, n.d.-b). These criteria constitute a major part of the World Heritage application materials assembled by the heritage nomination agencies. In the later development of heritage tourism, the skills and knowledge related to tulou construction and to tulou as a living space, unevenly solicited by various users, become rich resources for touristic narratives, presentation, and representation.

NOTES

1. Although round tulou are highly valued by experts, local residents may have different opinions. My informant Lin Fashan, a resident in the rectangular-shaped tulou Guangyu Building expressed that Guangyu Building, Fuyu Building, and Qingfu Building were more "complete" as they have outside gates. Although Zhencheng Building is a circular-shaped tulou, it is not as good as it doesn't have an outside gate. He also values the rectangular-shaped tulou for the exquisite decoration on the eaves that are not found on circular-shaped tulou.

2. World Heritage Convention, "Fujian *Tulou*," UNESCO, accessed May 1, 2014, http://whc.unesco.org/en/list/1113.

3. In this section, I use past tense for practices and beliefs that are no longer extant. Present tense is used for practices and beliefs that are still part of people's daily lives.

4. Feng shui is the art of geomancy originating from ancient China. It is particularly important in Chinese architectural tradition. Feng shui is used to select an auspicious site for dwellings by reference to astronomy and landscape features. Feng shui practices find correlations between humans and the universe by using instruments such as a magnetic compass. In addition to selecting a favorable construction site and orienting the building, people also practice feng shui in changing the positioning of the gate, hanging mirrors on the building, or placing certain auspicious items in the house to deflect negative energy. Feng shui is a common folk belief and practice among the Hakka people, but there are geographical variations. See Chen (2002), He and Luo (2000), and March (1968) for more detailed

discussions of feng shui and the relationship between people and their built environment.

5. Transcription of recorded interview conducted by the author in Chengqi Building on June 11, 2011.

6. Transcription of recorded interview conducted by the author in the office of the Yongding County Gazetteer Committee on June 28, 2012.

7. Door gods or threshold guardians are deities presented in the form of printed images of two ancient warriors. The images are pasted on doors to protect the residents from evil influences. See Koehn (1954) for an introduction to the Chinese door god.

8. See Phillips (2013) for further introduction of the kitchen god.

9. Feng shui beliefs and ceremonies are commonly practiced in house construction in this area. However, due to the larger investment of time, money, and human resources for the construction of tulou as massive communal dwellings, the ceremonies in such instances are usually more complicated and more seriously and strictly performed.

THREE

—ᜑ—

TULOU AS HOME AND
LIVED EXPERIENCE

TULOU, AS LIVING SPACES, incorporate not only the building's physical structure but also the local social structure, interpersonal relationships, and inhabitants' ideas, values, beliefs, daily activities, memories, and feelings. As evoked in the previous chapter in connection with construction customs and feng shui–based beliefs, residents have already associated symbols and meanings with their houses as crafted artifacts in geographical spaces during the process of construction. However, a tulou cannot be a home merely as a consequence of its physical form. Residents' customary daily routines and other social and cultural practices are fundamental in associating values and transforming the structure into a home that is central to the social lives and living world of village residents.

The process of constructing a home is a much longer and more complex matter than the work of building a physical house. It is a process of accumulating and embodying emotions, collective experiences, and memories through the personal and social lives of not only one generation but many. A significant part of the village residents' socialization processes, especially those expressed through rites of passage, is accomplished in tulou with the participation of other family or clan members. Tulou, as a common inheritance from the ancestors of the current residents, also convey messages from the past—making tangible the roots of the residents and enhancing their identities. Personal and shared experiences and memories make the physical artifact meaningful to the residents and are central in the creation of a sense of home and a sense of place. This chapter explores how tulou residents experience their communal dwelling practices and how tulou relate to local people's understanding of tradition, social relationships, and the meaning of the world and themselves.

LINEAGE RELATIONSHIP AND CEREMONIES

On a sultry summer evening, my host at Yucheng Building—Binghan's wife, Lizhen—invited me to Guangyu Building to attend a wedding with her. Knowing that I was interested in not only tulou but also people's lives in them, including this kind of ceremony, she offered to take me to the world that's usually not open to outsiders, the world referred to as "backstage" by Dean MacCannell (1973). In tourist settings, backstage is not meant to be presented to outsiders. Since Lizhen was a friend of the bride and needed to go to the new relocation village south of Hongkeng to participate in the ritual of accompanying the bride leaving her parents' home, she asked me to go to Guangyu Building first. I walked in the dark from Yucheng Building along the paved road on the bank of Jinfeng River toward the north end of the village. It was all quiet. The only sound was the occasional barking of dogs coming from the houses at the central part of the village when there were people walking by. In the darkness, the village was spotted with lights from tulou and other houses.

I got to Guangyu Building at around 10:00 p.m. There were several people sitting at the passage by the front gate of the building, enjoying the breeze and chatting with each other. One member in the group, a resident of Guangyu Building in his sixties who ran a small business of selling malt sugar candy, recognized me as we had met each other during my visit a year before. He greeted me and invited me to his house for tea. Knowing that I was staying at Yucheng Building, he said that my host Qiyin's grandfather was the brother of his grandfather. Both were the seventeenth generation of the Lin family. His grandfather was the youngest of the five brothers and was adopted by the family at Guangyu Building at the age of three. His ancestor created a "tooth tablet" engraved with family precepts, and it was kept by his granduncle Qiyin.

A couple days later, when I mentioned the tooth tablet to Qiyin, he said he would show it to me in the evening when he didn't have to be busily engaged in helping his son with the daily operation of the family hotel and restaurant. At around 8:00 p.m., Qiyin knocked on my door on the fourth floor and invited me to see the tablet. He led me to a big round table in the hall on the second floor. On the table was a big reddish-brown paper bag. Qiyin took the tooth tablet out from the bag and carefully wiped it with a piece of cloth. The dust on the tablet meant that it hadn't been taken out for a long time. In front of me, what was being referred to as a tooth tablet was a rectangular wood board painted in bright black of approximately twelve inches tall, four inches wide, and two inches thick. The most striking part of the board was the twenty-seven teeth encased with beaten gold (fig. 3.1). The teeth were arranged in six rows,

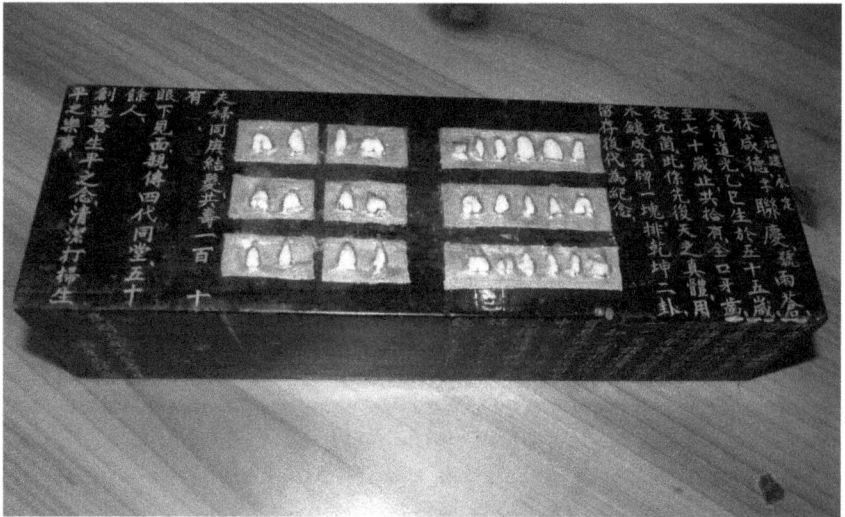

Figure 3.1   Tooth tablet. (Photo by author)

forming a hexagon that represented the heaven and the earth. Various precepts were engraved around the hexagonal pattern. Qiyin said the encased teeth represent the mouths of their ancestors.[1] It was like they were giving a sermon and teaching their descendants in person. It is a heritage object of Qiyin's great-grandfather Yucang from the beginning of the twentieth century. Qiyin regarded it as a family heirloom.

Qiyin's inheritance of the tooth tablet from his ancestor at Guangyu Building demonstrates the close relationship and genealogical ties between the residents of Yucheng Building and the residents of Guangyu Building. Therefore, my hosts in Yucheng Building were invited to participate in the ceremony of the bride's arrival at Guangyu Building. The ceremony was called the bride's arrival as the evening was not the actual wedding celebration but the auspicious time chosen by a feng shui master for the bride to enter the groom's house. Such auspicious times were usually in the evening. The actual celebration would be the next day. The central hall of the building had been cleaned. All the things stored or displayed in the central hall, including a traditional bridal sedan chair used for a touristic performance of the wedding ceremony, was moved either to the yard or to the passageways on the left and right sides of the central hall. The only furniture left in the central hall was a square wooden table with a censer on it (fig. 3.2). The groom's mother and several other close relatives in the clan family

Figure 3.2 The groom's father burning stick incense in the central hall of Guangyu Building as part of the ritual for the wedding ceremony. (Photo by author)

were busy cooking in the small kitchen. When Lizhen and I entered the building, the groom's father greeted us and seated us with the other guests. While waiting for the bride, we were sitting in the passage in front of the kitchen, drinking tea, eating snacks, and chatting with each other. At around 10:30 p.m., the loud sound of firecrackers informed everybody in the building and in the village of the bride's arrival. Several guests from the bride's side entered the house from the front gate, carrying symbolic items such as *zao* (Chinese dates, 枣), *huasheng* (peanuts, 花生), and *lianzi* (lotus seeds, 莲子) that together are homophonic to the phrase *zao sheng gui zi* (早生贵子), which means "bearing children soon" in Chinese. *Youdeng* (lamp, 油灯), which is homophonic to *ding*, the Chinese word for "family member," were also carried to express the same wishes. Like in many other places in China, these symbolic items are used in wedding ceremonies to allude to the bride's potential fertility (Yue 2019). The groom, dressed up in a new white shirt and black pants, had been waiting in the yard in front of the building, holding a big flower made of red silk in his hand. Upon her arrival, he led the bride by the hand to the ancestral hall. The bride wore a beautiful red dress, the traditional auspicious color for Chinese weddings. Standing shoulder to shoulder in the ancestral hall, the groom and the bride bowed to the ancestral shrine three times. Then, they turned around to face the front gate and bowed three times again.

The bowing is one of the most ritualistic sections that is still being practiced in the wedding ceremony. Local people nowadays tend to simplify the ceremonies in the rites of passage, except for funerals, when such simplification would be regarded as disrespectful to the deceased. Usually, the wedding ceremony lasts only one or two days, while a funeral can last from several days to more than two weeks. For weddings, it is increasingly common for people to hire a caterer or catering kitchen (*liudong jiujia*, 流动酒家) to cook for the family and guests. Some families will even invite guests to a restaurant for the ceremony. For funerals, the location for the ceremony must be at home, and the meals must be cooked by the host family and their kin in the clan.

After the bowing ritual, the groom led the bride to their bedroom on the third floor. The room, decorated with many balloons, was newly painted in white and had new plywood floors. On the bed, with brand-new red bedcovers and pillows, were two electric lamps that were used to represent the traditional kerosene lamps. A real kerosene lamp was burning on the nightstand beside the bed. As stated previously, the presence of the lamps expresses the family's wish for the couple to bring new members to the family after marriage. On the wall over the bed was a large papercut decoration of the character *xi* (happiness, 囍). On the opposite wall, above the TV stand, hung a picture of the groom dressed

Figure 3.3 Guests in the groom and bride's bedroom during the wedding ceremony at Guangyu Building. (Photo by author)

in a white suit and the bride in a white wedding dress. It had become popular in China for young couples, whether in the city or rural areas, to go to a studio to take wedding photos before their wedding day, with the bride in such pictures sometimes wearing a white "Western" dress.

The small room seemed to be especially crowded with all the furniture, decorative items, and guests (fig. 3.3). The bride, sitting on the bed, was happily talking with several girls who were her bridesmaids. The groom was chatting online with his sister who lived in the United States and was unable to attend his wedding. Some male companions of the groom were making tea by a small table. The crowd seemed to make the summer evening even hotter. The fast whirling standing fan didn't seem to help people cool off. Lizhen enthusiastically suggested that the bride and groom should install an air conditioner in the room.

Downstairs the groom's family was serving tea and preparing snacks for Ziwu and Binghan who drove the bride and those who accompanied her to the village. Two tables were soon set up at the central hall with food to treat the guests. The bride's younger brother, who was regarded as the most honored guest, was ushered to the seat closest to the shrine. The groom's family also placed all the other guests from the bride's side at the same table closer to the shrine to show their respect as those were seats of honor. After the feast, and several rounds of toasting, the ceremony ended with the hosts seeing off the guests to the sound of firecrackers.

TULOU AND LINEAGE SYSTEM

Lineage is central in defining the social relationship among the villagers. Like most other villages in the region, Hongkeng is a single-lineage community with members sharing the same surname. Since the Lin's first ancestors made their home in Hongkeng, the lineage has become more than twenty generations deep. The ancestral temple, as the symbolic center of the village and a symbol of the village patrilineage organization, enshrines the ancestor tablets recorded with names and marked with each generation. Genealogy and practices underpinned by kinship provide the frame for primary social relationships within village life.

Lineage bonds are sustained by written genealogies and, more importantly, by social interactions, such as obligations to offer help to villagers of closer relationship in the lineage when needed. Local lineage has a centripetal character. Although the written genealogies might be "retrospective constructions of the relations between lineages, making 'historical' sense of the present distribution of lineages on the ground" (Freedman 1971, 26), they are usually regarded by local people as historical truth through which people can trace their roots in history and build up connections in the present. People's relationships are often defined by their position in the lineage system. Those who have a closer relationship in the genealogy have more frequent social interactions. Ziwu's wife, Bilan, who is from another town, said, unlike her husband and her brothers-in-law, she was not very clear which tulou was owned by which branch except for the buildings of the eldest branch. She said, "I am not that familiar with the other people in our village except those in our own clan branch. For the families in our branch, I have visited them for ceremonies such as weddings and funerals and other important events." People belonging to the same lineage branch clean the ancestral tombs and give offerings to their common ancestors twice a year, once in the spring and once in the fall. For weddings and other celebrations, lineage branch members have to be invited to participate; it would be unusual for them to show up voluntarily at such celebrations. It is the hosting family's obligation or choice to invite certain people for such occasions. However, for funerals and unfortunate events such as an accident, lineage branch members are obligated to help or show sympathy even if they are not directly requested to do so. One afternoon when I was chatting with some residents at the side house of Zhengcheng Building, a woman in her sixties from another village came into the room with a gloomy face. The residents didn't know her, but she was introduced as having lineage connection with Hongkeng villagers. Her son had accidentally fallen down from a scaffold and was seriously injured. Being unable to afford the medical expenses, the woman came to ask help from the villagers who share the same ancestor with her. Because of the lineage ties, the residents of Zhencheng Building felt obligated to offer help and donate money to the poor lady. Such genealogical practices contribute to reinforce lineage bonds and solidarity. Even though there are differences and sometimes conflicts among the members, a lineage branch is an enduring group with many common interests and activities. Due to the tight lineage bonds, members of a lineage branch manifest a sense of in-group pride and identification that rests on the sum total of their wealth, prestige, and social influence. Even a humbler member of a lineage branch would proudly announce that the most celebrated tulou belongs to his clan branch even though he does not live in one of the buildings.

One of the older tulou in Hongkeng Village, Guangyu Building, was built in 1775 by Lin Fucheng, who was of the sixteenth generation of the Lin family in Hongkeng. He had five sons and eighteen grandsons. His grandsons brought him seventy-two great-grandsons. Lin Fucheng's descendants built seven tulou in Hongkeng Village, including Qingfu Building (built by the seventeenth generation of the Lin family in 1780), Kuiju Building (built by the eighteenth generation in 1834), Yucheng Building (built by the nineteenth generation in 1885), Zhencheng Building (built by the twenty-first generation in 1912), and Qingcheng Building (built by the twenty-first generation in 1938). As the lineage expanded and the family grew, more tulou were constructed to accommodate the new lineage and family members. The buildings are the reified form of the lineage relationship between the tulou residents. The knowledge of which generation built which tulou circulates within the lineage branch and plays an important role in the lineage members' sense of place and community identification.

The lineage network became the focal point whenever I asked my tulou residents about their family history. Fashan, who is a senior member of the eldest lineage branch living in Guangyu Building, is proud that seven other tulou, including the three most famous buildings in the village, were built by ancestors of the residents who live in his tulou. He claimed that Guangyu Building should be regarded as the origin of those other buildings. My conversation with him about the relationship between Guangyu Building and some other tulou illustrates how the tulou become physical manifestations of lineage development: "Fucheng built Guangyu Building. He had five sons. His fifth son, who was the seventeenth generation, built Qingfu Building. Kuiju Building was built by a member of the eighteenth generation from Fucheng's family. Yucheng Building was built by the nineteenth generation. Fuyu Building was built by the twentieth generation. Zhencheng Building was built by the twenty-first generation. And Qingcheng Building was also built by Fucheng's descendants."

In Hongkeng, the distribution of lineage branches can be mapped by the locations of the villagers' communal dwellings.[2] This reflects the importance to the village of both the buildings and the social groups, and it speaks to the way tulou represent a distinctive aspect of the more widespread social institution of village lineages in Southeast China. Tulou is a significant part of the local social ordering system, one linked to lineage grouping and development. When I asked Ziwu about his family history and what he knew about his ancestors, like Fashan and many of my other informants, he naturally incorporated tulou into his narrative:

> There are five lineage branches in our village. The branches are descendants of five brothers who are the eleventh generation [of the Lin family in Hongkeng]. I belong to the eldest branch. My ancestor is Binggong. People from Zhencheng Building, Kuiju Building, Fuyu Building, Guangyu Building, and Qingcheng Building have the same ancestor as

I do. Actually, there is the old eldest branch and the new eldest branch. The old eldest branch is constituted of all descendants of Binggong. And members of the new eldest branch share Fucheng as their most direct common ancestor. Now all tulou selected for tourism development belong to the new eldest branch. Fucheng built Guangyu Building. Those who live in other selected tulou are built by descendants of people who lived in Guangyu Building.

Fucheng is the sixteenth generation. My house, Yucheng Building, was built by our ancestors of the eighteenth and nineteenth generations. It was built cooperatively by an uncle and his two nephews. One of the nephews, Shanqing, is my direct ancestor. He is the nineteenth generation. The uncle Xiangyi is Bingxiang's direct ancestor of the eighteenth generation. I am the twenty-fifth generation. At that time, the nephews were rich. So, the uncle took the lead and built the house with his nephews. Actually, the house belongs to three big families. The members of one family went overseas, so now only two big families are living in the building. My great grandfather also went overseas. He never came back home, so my great-grandmother lived like a widow. She later adopted a son, who is my grandfather. My grandfather was a gambler, and he sold most of our rooms in this building for gambling. My father later bought those rooms back one by one. My father was a carpenter and he worked in different places until his midlife, when he came back home for farming.[3]

As illustrated by narrative histories of this type, members of the eldest branch feel proud because their branch is one of the biggest and most prosperous. The size and power of the lineage branches are not equally developed. The youngest branch has the largest population, followed by the eldest branch. The populations of the second and third branch are in the third and fourth places. Ziwu informed me that the fourth branch is facing the danger of disappearance, "Many generations of the fourth branch had only one son in their patrilineal line of descent. The men of the current generation married into, and live with, his wife's family." The larger and wealthier the lineage branch is, the more tulou it owns. Thus, the number of tulou, especially those so-called special tulou, represents the prosperity and power of a lineage branch.

Geographically, the village is divided into what the villagers call upper village and lower village, with the two halves diverging at the spot where Lin's Ancestral Temple is located. Both parts are further divided into five or six smaller social-geographical units. The five units in the upper village are Airenxia, Wulouxia, Cuntou, Shangkuliao, and Dunzi. There are six units in the lower village: Zhongjie, Xibei, Gongwangba, Zhengcheng Building, Xiawu, and Shuiwei.

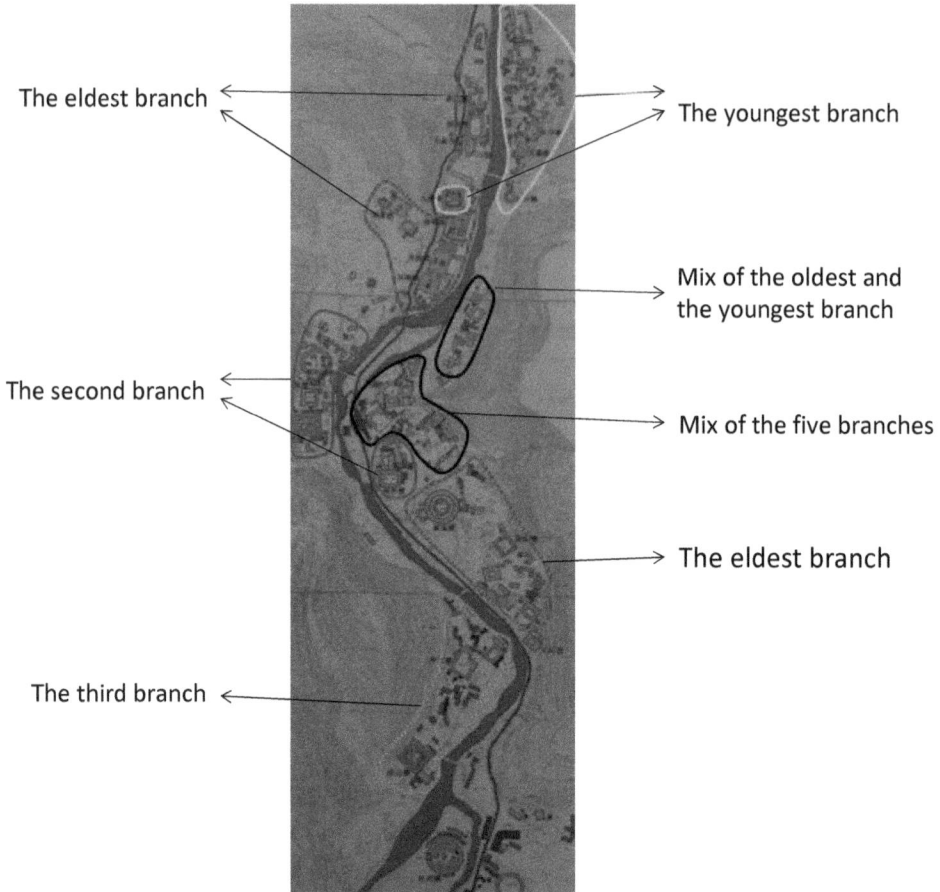

Map 3.1. Location of the five lineage branches in Hongkeng Village. The location of lineage branches was identified with help from Ziwu, a local resident and the vice-chairman of Hongkeng Villagers' Committee. (Map created by author based on field survey, interview, and material provided by the Archive of Yongding Office of Cultural Heritage)

In general, the division of the units links to the genealogy system of the Lin lineage. Most residents of the same branch live in relatively concentrated areas (map 3.1). Villagers form a clear mental map of the lineage distribution through daily interactions and lineage activities as well as the locations of tulou. Zhiwen pointed out the locations of each branch on the village map I showed him: "Most members of the eldest branch live at Baduipian, Xiawu, and Zhengcheng

Building. Some live at Huangzhuxia and Zhongjie. The majority of the young-est branch live at Cuntou with a few living at Baduipian, Huangzhuxia, and Zhongjie. Some of their members are rich people doing business in other places. The second branch mainly lives at Xibei and Gongwangba with some members living at Zhongjie. The third branch lives at Shuiwei and Zhongjie. Zhongjie is a place with members of all five branches living there." The association of the lineage branches with the geographical units is related to the expansion of the Lin lineage and its development through history. And this contributes to local people's identification and sense of place.

Gongwangba is in the middle of the village on the east side of the riverbank between Lin's Ancestral Temple and the Zhengcheng Building. There are sev-eral tulou ruins in this vicinity. It is the first settlement area of Lin's ancestors. After walking through several old tulou and ruins of tulou, I encountered a villager on the narrow path between the tulou who pointed me to the location where his ancestors first settled and the first tulou was built. The ruin now is a vegetable garden, and only the stone foundation left on the ground shows that it used to be a tulou. The greatest number of tulou ruins is found in this section, and all of them have been turned into vegetable gardens. Mixed populations of all five branches live in this earliest section of settlement for Hongkeng villag-ers' ancestors. As stated, the eldest branch and the youngest branch have the largest population, and they are also the most prosperous. The eldest branch possesses most of the well-preserved large tulou on the north end of the village along the west side of the riverbank and in the area around Zhencheng Build-ing. People of the second branch live collectively in the middle of the village on the west side of the riverbank. Most of the third branch families live at the south end of the village. The youngest branch developed tulou sites at the north end of the village along the east side of the riverbank. I was told that, when the ancestor of the youngest branch was born, all of the farming land was already distributed to his four elder brothers. So, he got land on the hills surrounding the village. He and his descendants planted tea on the hills and produced tea oil, which they sold far and near.

The map of lineage branch locations demonstrates how the number of tulou grew in the village as the lineage expanded. The lineage ancestors used to live in the same house and shared a single roof. When the family grew, the lineage developed its lineage branches, which built separate tulou in different places in the village. As a particular branch grew, the new tulou would be built close by, so that a branch stayed together geographically. Tulou generally reify the social organization and structure of the local community while reinforcing lineage solidarity and personal identity. The architecture is also a manifestation of the uneven development of the different families and lineage branches as well as the hierarchical structure in the lineage-based social network. Lineage does not resemble a homogeneous unit. The disparity in accumulated resources and power through generations increases the gap between the different branches, which can be reflected by the number and size of tulou cooperatively constructed by the lineage members. In the current development of heritage tourism, tulou, as the inherited property from the inhabitants' ancestors, become an important factor in determining the cultural representation, economic resources, and social status in the local community.

LINEAGE-CENTERED SPACE DESIGN

As mentioned in chapter 3, there is a spatial distinction in tulou in terms of sacred activities and the secular world. The central hall, which is the most distinctive section of the tulou and is shared by all the inhabitants, is regarded as a symbol of family and lineage solidarity. The central hall or ancestral hall is designed to occupy the central or rear space on the axis of the building, oftentimes at a higher position. At the back wall of the hall a shallow hole is dug to enshrine ancestral tablets or gods such as Guanyin and Yanggong. Sometimes people simply put the deity on a table against the wall. The specially designed space serves the local centripetal lineage and family structure, promoting its cohesive quality. The allotment of housing is based on the number of sons in a family. All households in the building regard the central hall as their shared spiritual center, a sacred space for ancestor worship and for specific ceremonies. Residents' rites of passage from birth to death and all ritual practice during festivals from spring to winter, as well as other ceremonies and public events, are all related to this space with great and multifaceted symbolic meanings to tulou residents.

The central hall and the entrance hall function separately as sacred and secular public spaces. While announcements and notices, such as residential financing information, that do not relate to ceremonial uses or sacred themes are put on the walls in the front hall, the transitional periods of life, such as birth, naming, marriage, funerals, and festivals, are announced in written form on the walls of the central hall. The central hall is of great personal and social significance even when people no longer live in it. In otherwise collapsed and uninhabited tulou, one can sometimes find well-maintained central halls with burnt incense and new couplets. Even when the building is in ruins or turned into vegetable gardens, the central hall may still be carefully preserved as a sacred center for an extended lineage family when possible (fig. 3.4).

Figure 3.4 The central hall of a tulou building right outside of Hongkeng
Village with the surrounding structure collapsed and turned to vegetable garden.
(Photo by author)

In the past three decades, newer single-family housing with modern facilities has become more popular in the region. Many residents have moved
out and built cement houses or live in other locales for jobs or business. Even
though such residents keep their allotted rooms padlocked and largely un-
used, many of them still have ceremonies in their tulou for occasions such as
weddings, funerals, and times of ancestor worship. During festivals, especially
the Chinese New Year, former residents come back to their tulou and put red
couplets on their kitchen door.[4] With the current tulou residents, those who
move out also donate money to buy incense and firecrackers for the common
celebration of the Chinese New Year and for maintenance of the building.
Although they no longer physically live in the tulou, their connection with the
building continues in the form of doing important ceremonies in the building
or simply changing the couplets on the kitchen door each year.

THE DECLINING POPULARITY OF TULOU
AS RESIDENTIAL DWELLING

Even though tulou are celebrated as World Heritage Sites, tulou are not considered by local people as meeting people's living needs or standards in contemporary society. People complain that the rooms are too small and not bright. And
it is not convenient as there are no modern bathrooms in tulou. The communal
living style is no longer suitable for people, especially the younger generation,
who desire and value private space for themselves and their nuclear family.
Thus, people gradually move out of tulou or renovate tulou to improve their
living environment.

When I visited Shizhong Township, which is on the border with Yongding
County, the big tulou I visited had only one or two households still living there.
The buildings were not well maintained. There was litter in the deserted rooms
and a layer of dusts upstairs along with mold on the first floor in the sections
that were not inhabited. The same is true for the buildings in Hongkeng that
are not designated as protected heritage sites or are not engaged in touristic
activities. An example is Zhengdong Building behind Yucheng Building. It
is a rectangular-shaped tulou that is further away from the tourist route. The
residents of the building moved out to single-family homes they built in the
1980s. The empty building is now used to raise chickens.

In the 1980s and 1990s, people started to regard tulou as a symbol of under-
development. Before the development of heritage tourism, local people were
not satisfied with the condition of tulou. Those who married villagers who

lived in tulou often expressed feeling scared in the evening while sleeping in the depopulated tulou. My host Jianwu's wife told me her impression of tulou and experience of living in it before the family turned their residential home into a commercial space.

> Tulou used to be very dirty. People were afraid of going into Fuyu Building, as it was deep and dark. There were coffins in the building. In the past people prepared their coffins when they were in their sixties. People were afraid of the coffins. Our building was the same as other buildings. Before our marriage, my husband invited me to visit his house. There was a wall and a tobacco flue-curing house in front of the building. The building was very dark. There used to be a small house in the court-yard, and it was torn down recently. The walls of the central hall used to be dirty black. Later my husband painted it in white. The rooms next to the stair and the central hall used to be a pigsty. There were also pigs raised in the courtyard. It was very dirty. It looks better after my husband repainted it.
>
> In the past, many rooms upstairs were locked. A unit on the other side of the building was owned by another family. They have a closer relation-ship with Binghan's family. As their family were well educated and got jobs elsewhere, they gave their property to Binghan. For Binghan's family, there used to be only two elders living in the building. All the young people left the village to become workers in the cities. They later came back to the building to run the restaurant and hotel business after seeing my family making money by selling rice wine.
>
> In the past, this room by the stairs and the room for smoking rice wine at the back were used as a pig pen. We also raised pigs at what is now a cen-ter yard. It was dirty. My husband was a migrant worker. We didn't have a phone or cell phone. It was hard for me to contact him. Sometimes I didn't receive my husband's letter for half a month. I stayed here by myself and didn't go back to my own parents' place. I couldn't go anywhere as I didn't have money. When my parents-in-law sold their pigs, they gave me twenty yuan. When they sold tobacco leaves, they gave me thirty yuan. That's all the money I got for the whole year. I asked my husband to take me with him. He said he needed to check whether there's a place for both of us to stay at the place where he worked. He would stay there alone if there was no place to accommodate both of us. In the evening, I was too scared to sleep by myself. I asked the little girl in Binghan's family to sleep with me.

Table 3.1. Occupancy rate of the major tulou buildings in Hongkeng Village in 2007. (Statistics provided by the former director of the Office of Tulou World Heritage Nomination.)

Building Name	Number of Households	Number of Residents	Number of Rooms	Current Inhabitants
Zhencheng Building	19	105	178	70
Fuyu Building	20	95	182	45
Kuiju Building	25	116	122	69
Fuxing Building	16	69	78	71
Qingcheng Building	2	11	80	7
Guangyu Building	25	114	94	57
Rusheng Building	12	62	48	25
Total	**119**	**572**	**782**	**344**

The memory of tulou being a place that is not clean and that can make residents feel scared is shared by Binghan's wife, Lizhen. Tulou residents' lives were hard with limited resources. Despite the glorifying narrative of tulou in tourism, in private conversations there's a negative sentiment in people's memory of the image of tulou and their lives in tulou in the past.

For a period of time before tulou were nominated as UNESCO World Heritage Sites, the population of tulou residents had greatly decreased, as many residents either moved out and built new houses or became migrant workers. In many cases, only a few old people stayed in the tulou. Since the end of the 1970s, people started to build small, rammed-earth houses and later brick-and-cement houses for single-family use. Those who still live in tulou were usually those who could not afford to build their own single-family house. At the same time, like many other rural communities in China, young people left the village for job opportunities in more developed regions. In 2007, a year before the UNESCO World Heritage nomination, the local government measured the number of residents in the major tulou in Hongkeng Village with statistical data (table 3.1). In most buildings, such as Fuyu Building, Kuiju Building, Guangyu Building, and Rusheng Building, only half of the residents still lived in the building, with the other half having left their tulou as migrant workers, college students, or individuals moving to single-family houses. One of the oldest tulou, Wansheng Building, has only two elder residents.

CONCLUSION

As demonstrated in this chapter, tulou constitute an important part of the local social and political landscape of southwest Fujian Province. The tangible
form of tulou maps, mediates, and contests local social relationships as well
as local social and cultural life. At the same time, social and cultural practices
construct tulou residents' sense of home and social space. What transforms the
physical structure into a home is "a routine set of practices" and "a repetition of
habitual social interactions" (Nowicka 2007, 72). In Hongkeng Village, lineage
is the fundamental social structural unit in which historical elements, including narratives and a sense of identity, are transmitted to the present time so as
to create social bonds and to determine local social relationships. Tulou as a
physical demonstration of the local lineage system provides people with stable
and conservative objects to understand their historical roots and their present
position in the network of social space.

However, forming a single lineage in one village territory does not mean that
society is homogeneous there. The accumulated resources and power of certain
branches within the lineage system, in this case, represented in the form of the
large architectural structures, amplifies the imbalance of the development of
the lineage branches. Despite the developmental imbalance, within the lineage
system there is a strong sense of kinship and connection among the villagers.
The lineage structure perpetuates concomitant relationships over time and,
in some cases, over space. The villagers identify with each other because they
share common ancestors and memories of ancestors reified in the form of the
family temple, tulou, and genealogy narratives as well as ancestral rituals and
ceremonies.

Both tulou and people's lives in tulou have evolved. Some traditions are
continued, such as the important ritual of bowing to ancestors and using symbolic objects in wedding ceremonies to express wishes for new family members.
Some traditions are emerging, such as shooting wedding photos at a studio
or hiring caterers to prepare the wedding feast. While funerals are held in ac-
cord with a lot of local traditions, sending wreathes and hiring a professional
band with modern instruments have become a new tradition in the funeral
ceremony. Tulou are also modified to meet people's needs. If people want a
room to be brighter, they might expand a small window on an upper floor or
open new windows on a lower floor. They might paint the rooms or change floor
materials. Before tulou were listed as UNESCO World Heritage and Hongkeng
Village were designated as a World Heritage Site, people would either seek new

living spaces or renovate tulou in the way they would like them to be when they found tulou did not meet their daily needs and living standards.

The features of tulou that are valued are still valid, but as society develops and circumstances change, tulou, along with many other traditional houses, are no longer desired by people, especially the younger generation. Sharing living space means limited privacy and more frequent interpersonal interactions, which may sometimes lead to disputes and conflicts. After the tulou World Heritage nomination, the status and values of tulou are subject to change. The transvaluation goes both ways. In some places, such as Shizhong Township right outside the border of Yongding County, tulou have been falling into the state of what Babara Kirshenblatt-Gimblett (1995, 369) described as "the obsolete, the mistake, the outmoded, the dead, and the defunct." Even in Hongkeng Village, some tulou, especially those that are not on the tourist route, are deserted, broken, and dilapidated. Some are turned into sheds for livestock. For the tulou designated as heritage objects, the transvaluation goes a different direction. As described in the following chapters, they start to attract residents with new functions and values.

NOTES

1. On the introduction section of the tooth tablet, it says that the teeth were collected from Lin Xiande, who was born in 1845. And the collection period started when he was fifty-five years old and lasted until he was seventy years old. He and his wife lived with three generations of descendants who formed a household of over fifty people.

2. The Chinese concept of clan here is distinct from other anthropological concepts of clan. The key factors to note are that it is nontotemic; it is based on actual known relations; and it is generally exogamous. The clan is lineage based.

3. Transcription of interview conducted at Yucheng Building on June 9, 2011.

4. Red couplets, or simply couplets, are a folk art in China for wall decor. They are usually two lines of rhymed and poetic words expressing auspicious wishes to the family written or printed in calligraphy with black ink on vertical strips of red paper. They are usually hung on each side of the gates or other doors in a house. For residents in tulou, each pair of couplets on the kitchen door represents a household in the building.

FOUR

—⟋⟍—

WORLD HERITAGE NOMINATION AND THE INSTITUTIONALIZATION OF TULOU

TULOU HAVE BEEN RESIDENTIAL dwellings in Fujian since the twelfth century. However, the form was "hidden" in the once remote mountainous region of Southeast China for most of history. Not until the twenty-first century, when it was no longer being constructed, did it start to have a new significance. On July 7, 2008, following a ten-year nomination process, local people in Fujian finally received the "good news" from Quebec, Canada (where the thirty-second session of the UNESCO World Heritage Committee was meeting): "representative" tulou would be given the name Fujian Tulou and recognized with the UNESCO World Cultural Heritage designation. With this designation, the tulou houses, the dwellings of generations of Hakka and Minnan people in the area, began to develop a new identity. The announcement became headline news for many local and national media outlets, which was only the start of the continuous process of publicizing and promoting these large, multistory, multifamily vernacular buildings enclosed by thick rammed-earth walls and hidden in the once remote and isolated valleys of this mountainous Chinese region. Following the designation, various public agencies and private firms flocked to this high-profile heritage space. The entrepreneurs and public-private corporations brought capital investments, while the public agencies provided management structures, policy determinations, promotion efforts, and scholarly interpretations. Tulou, locally constructed private dwellings, were transformed, becoming symbols of national heritage and public property to be shared with the world.

In this chapter, I focus on the transformation of tulou from residential homes to UNESCO World Heritage and situate the examination of tulou

within a larger context, namely, the noticeable national interests in internationally recognized/legitimated World Heritage and international efforts to
preserve heritage across the world. Tulou heritage nomination was an elite-led
but local effort to enter the global cultural and institutional system in a way
that is consistent with the national pursuit of globalization and modernization. In the past two decades, massive endeavors have been undertaken in the
study, nomination, preservation, promotion, and commodification of heritage
in cultural, political, and economic spheres across China. Fujian Tulou World
Heritage, like many other internationally or nationally recognized expressions
of heritage, is part of the ongoing global heritage process and discourse. In
this process and discourse, UNESCO, as an international organization that
deals with the safeguarding and listing of World Heritage Sites, possesses the
authority to inscribe and legitimate a certain object or cultural expression as
heritage. Taking tulou as a specific case that reflects China's heritage discourse
and practice on the regional level, this chapter explores questions such as: How
did tulou, as residential dwellings in a remote mountainous area that were not
known to the outside world, receive national attention in the first place, and
then how did they gain international recognition? What is the on-the-ground
process of making heritage in specific communities? Why, in the Fujian Tulou
case, did the local government want tulou to be nominated as a UNESCO
World Heritage?

The nomination making is a process of exercising power and creating meaning, and
as previously stated, governments play a significant role in the current context
of the heritage movement in China. As a centralized regime that takes economic development as the central priority in its pursuit of modernization and
globalization, the government articulates a top-down policy toward heritage
in which national and local authorities play critical roles while local residents
usually have limited agency in the nomination process. The cultural, political,
and financial resources as well as the discursive power possessed by government and cultural experts enable them to become authoritative forces in the
heritage-production process. As such, local authorities' understanding of heritage discourse and practice has vital impacts on the nomination practice and
on the local communities.

The process of tulou World Heritage nomination was a massive political
and cultural undertaking that had transformative impacts on tulou and local
communities. Through the process, local government agents had to learn and
embody a system of international and national heritage regulations and other
heritage-related knowledge. It is also a process through which local government agencies and communities reevaluated tulou and local culture, engaging

in the metacultural practices that resulted in new understandings of tradition and culture as heritage. The terms UNESCO used to describe and evaluate World Heritage objects, such as *masterpiece, human creative genius, outstanding example, outstanding universal significance,* and *aesthetic importance,* were etic (comparative) terms local people did not use in their daily lives. In the process of World Heritage nomination and tourism development after the designation, local residents frequently encountered these words in their lives, mainly through the promotional work conducted by experts and authorities. The institutional practice of securing the tulou UNESCO World Heritage nomination not only introduced an entirely new vocabulary and set of concepts into local people's daily lives but also transformed their conception and perception of the houses that their ancestors built.

THE "DISCOVERY" AND "REDISCOVERY" OF TULOU

In narratives related to the tulou World Heritage nomination, tulou's value was "discovered" not by the inhabitants but by outsiders. Despite the giant structures' significance in local communities that is promoted in the current heritage discourse and practice, tulou, hidden in a once remote area seldom visited by outsiders due to the rugged mountain paths that led to them, were unknown to the outside world for many centuries. How did the residential houses in the valleys become well known to the world? The story starts with the recent history of the dwellings in the valleys of western Fujian, which gained fame in the field of architecture studies in the second half of the twentieth century. Some architecture scholars were the first outsiders to seek out tulou, which is not a surprise due to the magnificence of the buildings. I would say it is hard for architecture scholars not to pay particular interest to tulou due to their materiality, as presented in the previous chapter. In the 1950s, a local student from Yongding, who went to study architecture in Nanjing, a large city in eastern China, introduced tulou to his professor, Liu Dunzhen. Afterward, Liu selected two tulou to represent Hakka vernacular architecture in the teaching material he edited. He also published an article on the study of tulou. His study was later introduced to architecture scholars in other countries such as Japan and the United States. Some scholars were fascinated by his description of tulou and decided to make a trip to study tulou in person.[1] Thus, tulou started to receive domestic and international visitors who were primarily architectural specialists.

The narrative that the fame of tulou emerged due to their unique architectural style piquing the interest of architecture scholars seems only to circulate among the local intellectuals. The much more widely circulated narrative is much different. Local people often told me and other tourists that it was outsiders (more specifically, foreigners) who made tulou popular. Local opinion often attributes the popularity of tulou to a particular political circumstance involving the United States. The widely transmitted anecdote is that during the Cold War period, an American satellite spotted tulou and misidentified them as China's missile launch base, which alerted the American military, who reported it to the White House. Later on, the American government sent people to Fujian to investigate secretly, and they found that the massive structures were private residential dwellings rather than military facilities. Another version of this local anecdote says it was not until 1972, when US president Richard Nixon made his famous visit to China, that the Americans determined the purpose of those missile launch base–like structures. The local people seem convinced of the story, especially with its wide use in promoting tulou and heritage tourism. A local tour agency even prints the story on its advertising brochures. In 2012, I asked local scholar Su Ziqiang about the story. In his office at the Yongding Overseas Chinese Association (Yongding xian huaqiao lianhehui, 永定县华侨联合会), he provided me with a version of the narratives that sounds more historically credible:

> It is true that the United States satellite detected tulou. At the end of 1960s, the US satellite found some massive structures with a round shape in southwestern Fujian. Probably because of technological limitation of satellites at that time, the United States couldn't figure out what the structures hidden in the mountains were. In 1970, President Nixon sent a vice president whose name was [Spiro] Agnew to Taiwan and informed Taiwan that there was something like a missile launch base in southwestern Fujian. He suggested Taiwan do some investigation of it. There happened to be a man whose ancestral home was in Yongding serving in the military in Taiwan. He was sent to Yongding to do the investigation. To the relief of United States Central Intelligence Agency, he returned to Taiwan and reported that it was a residential house rather than a missile launch base. The man's name is Fan Jingmin. He is still alive and is now a councilman of Taiwan Yongding Fellow Association. I have met him in Taiwan.

As noted by Su, the narrative of the American satellite's discovery of tulou and relating tulou to international relationships in a historically sensitive period lends an entertaining and legendary tone to the story. But how much did it actually contribute to the popularity of tulou? While the story is regularly circulated with various versions and repeated by all kinds of agencies, the materiality is regarded as fundamental for tulou to gain its fame nationally and internationally. The story added vividness to the promotion of tulou. However, Su emphasized that it is the uniqueness of tulou that makes it known to the outside world. In other words, the materiality of tulou is regarded as fundamental to its fame:

> It is rare to see this kind of big building with dozens of households living in it. Architecture experts, including those from Japan and the United States, were very interested in it. They publicized tulou to the outsiders because it is unique. They were impressed by the rich culture and the residential model of a large lineage family living together. Tulou is representative of East Asian culture. Thus, the foreign experts regard it as something of great value. At that time Chinese, especially Yongding local people, did not think there was anything special about it. We live in tulou every day. We didn't feel it was special. However, for the outsider, they examined tulou from the perspectives of culture and anthropology, and they thought it was an extraordinary construction and creation by human beings.

As Su pointed out, although tulou became famous as unique architecture, local people did not regard their residential houses as particularly unique.

When I asked Hongkeng villagers what they thought about tulou, the most common answer was that they didn't think of tulou as "outstanding" and "exceptional," as the UNESCO web page described it, before it was designated as a World Heritage Site. They said they lived in tulou every day. It was part of their daily life that did not stand out from their habitus. At the same time, tulou was not incidentally discovered by a foreign satellite and then suddenly became institutionally recognized. The transformation of tulou began with the intersection of local agencies' political and cultural practices and the interests of outsiders in the structures.

Local elites, such as Su, currently play an important role in local advocacy for tulou. Su, who is in his sixties, has been studying Hakka culture for a long time. He grew up in a tulou and later came to work in the county city as a government official as well as a local intellectual. In addition to being chair of the Yongding Overseas Chinese Association, he is also chair of the Yongding Hakka Friendship Association (Yongding kejia lianyi hui, 永定客家联谊会) and deputy chair of the Yongding Hakka Tulou Culture Research Association (Yongding kejia tulou wenhua yanjiu hui, 永定客家土楼文化研究会). Even though he is a local expert on Hakka culture, like other local residents, he admitted that he did not pay special attention to tulou when he was living in it. He said, "We grew up and lived in the tulou every day, so we became habitually accustomed to it."

Su attributed his initial interest in Hakka culture and tulou to the opportunity to be in contact with outsiders due to his job in the county government. He was in charge of the international communication sector in the Yongding County Department of Publicity. His job granted him the opportunity to have direct contact with experts and journalists who came to visit tulou. This contact with outsiders who showed special interest in tulou made him realize that his residential home was "unique and spectacular" in terms of architectural form and cultural meanings. That motivated him to collect related materials and to interview local elders about tulou. He said he "gained a better understanding of tulou and Hakka culture" in the process and the knowledge also served his communication work in the county department of publicity:

> At the beginning, we collected information about the architectural form, history, and cultural meanings of tulou. For instance, we studied how the big tulou with more than a hundred rooms were built. There are so many tulou in Yongding that even though we are local, we couldn't get a full understanding of them. There are many tulou that we don't have enough time and energy to study. The only thing we can do is to study and promote several of the most representative tulou so that the outside world can get a sense of what a tulou is.

This statement reflects how knowledge of tulou started to be collected and institutionalized on the ground by local people.

Local agents such as Su play a significant role in the defamiliarization (Foster 2011) and transvaluation (Kirshenblatt-Gimblett 1998a) process of tulou. They are both local intellectuals and associated with local government through the various titles that they hold in the governmental institutes. They were the first people from the local community to consciously and reflectively study their communal residential houses. Their expertise as well as their positions authorize their tulou-related opinion and practice both in local community and in outsiders' reconsideration of local culture. Their study of tulou and their promotion of tulou with a high level of cultural awareness greatly contributes to tulou heritage making, becoming another major representative feature of Yongding County in publicity work aimed at promoting its local image to the outside world. At the same time, such local agents also work on advocating and promoting the value of tulou in the local community. This led to the process of more local people starting to defamiliarize their residential home and reevaluate tulou through local scholars' and outsiders' eyes while exploring new opportunities through the "discovery" of tulou as something of great value besides its residential function. This process was greatly accelerated and intensified after tulou was designated by UNESCO as World Cultural Heritage.

THE LONG JOURNEY OF THE NOMINATION PROCESS

When UNESCO placed Fujian Tulou on the UNESCO World Heritage List, its web page described them as "exceptional examples of a building tradition and function exemplifying a particular type of communal living and defensive organization" (UNESCO, n.d.-c). As stated previously, tulou were originally constructed in the southwest mountain region of Fujian Province by local people as large community residences (usually for a large extended family or lineage group) and, in some cases, for defense. Since 2008, *Fujian Tulou* has become the legitimated and official term used by UNESCO and governments to refer to the designated "featured tulou" and tulou clusters. These tulou, according to UNESCO's assessment, are "the most representative and best preserved." The UNESCO web page explains that "Fujian Tulou is a property of 46 buildings constructed between the 15th and 20th centuries over 120 km in south-west [*sic*] of Fujian Province, inland from the Taiwan Strait" (UNESCO, n.d.-c). Although *Fujian Tulou* as a term is not commonly used in local people's daily life, the appearance of the term on the UNESCO World Heritage List officially and institutionally separates those universally recognized massive structures and locations apart from the rest of the regional landscape as common world property and heritage sites. In the following, I take Hakka communities in Yongding County as an example to reflect on and analyze the on-the-ground process of heritage nomination at the local level.

It took ten years of effort and nomination work for tulou to become the thirty-sixth expression of World Heritage in China and the second in Fujian Province. The other one in Fujian is Mout Wuyi in the northern part of the province, which is recognized as a UNESCO World Cultural and Natural Heritage Site. According to local officials' accounts, in a sense, the enlisting of Mount Wuyi triggered local desire for the nomination of tulou. I was informed that it was during the nomination of Mount Wuyi in the late 1990s that some experts from UNESCO and from national institutes who visited Fujian told local officials that tulou would be eligible for World Heritage List nomination.

However, at that time local officials and experts had only a very vague conception of what the World Cultural Heritage designation meant and knew nothing about the nomination process. Zhen Dinglai, an official in charge of the nomination work in Yongding County, described to me the mixed feelings of excitement and uncertainty: "The experts and scholars told us that tulou also met the standards for World Cultural Heritage nomination. We were excited to hear that. We didn't know we had a treasure. But at that time, we were confused. What is World Cultural Heritage? We had no clue at all." The excitement came from confirmation from the institutional and cultural authorities of their possession of a "treasure" of high value, and the uncertainty stemmed from a lack of experience of and knowledge about the World Heritage concept. Therefore, the local government actively sought the aid and expert opinion of outsiders. Officials started to familiarize themselves with the regulations and systems of the UNESCO World Heritage List nomination process. Scholars and experts were invited by the local government to study tulou and offer instruction on heritage nomination. Local officials also visited Mount Wuyi, Pingyao Ancient City, and other UNESCO World Heritage Sites to learn from their experience. Gradually, they absorbed and embodied, wholly or partially, the value, knowledge, and operational system of formal heritage and World Heritage List nominations from authorized and experienced personnel, institutes, and organizations.

The first step of the nomination preparation work for the local government was to work with related experts and identify whether tulou matched the UNESCO World Heritage selection criteria. Ten selection criteria are listed on the UNESCO web page, six for cultural heritage and four for natural heritage (UNESCO, n.d.-b):

1. to represent a masterpiece of human creative genius;
2. to exhibit an important interchange of human values, over a span of time or within a cultural area of the world, on developments in architecture or technology, monumental arts, town-planning or landscape design;
3. to bear a unique or at least exceptional testimony to a cultural tradition or to a civilization which is living or which has disappeared;
4. to be an outstanding example of a type of building, architectural or technological ensemble or landscape which illustrates (a) significant stage(s) in human history;
5. to be an outstanding example of a traditional human settlement, land-use, or sea-use which is representative of a culture (or cultures),

or human interaction with the environment especially when it has become vulnerable under the impact of irreversible change;

6. to be directly or tangibly associated with events or living traditions, with ideas, or with beliefs, with artistic and literary works of outstanding universal significance. (The Committee considers that this criterion should preferably be used in conjunction with other criteria);

7. to contain superlative natural phenomena or areas of exceptional natural beauty and aesthetic importance;

8. to be outstanding examples representing major stages of earth's history, including the record of life, significant on-going geological processes in the development of landforms, or significant geomorphic or physiographic features;

9. to be outstanding examples representing significant on-going ecological and biological processes in the evolution and development of terrestrial, fresh water, coastal and marine ecosystems and communities of plants and animals;

10. to contain the most important and significant natural habitats for in-situ conservation of biological diversity, including those containing threatened species of outstanding universal value from the point of view of science or conservation.

Some terms in the criteria are defining features of what are eligible for the nomination, including *architecture* or *technology, monumental arts*, and *town-planning* or *landscape design*. Meanwhile, the criteria also contain many value-related descriptive words and phrases that are hard to measure with an objective standard, such as *masterpiece, unique testimony, outstanding example*, and *universal significance*. In these cases, authorized experts' specialized knowledge and personal judgments are vital in the evaluation of nominated sites.

UNESCO requires that nominated sites be of "outstanding universal value" and meet at least one of the ten selection criteria. These criteria not only are the main "working tool" for UNESCO staff but also offer significant referential information for local governments to evaluate whether a site has the possibility of being nominated and what they need to do to meet the nomination requirements. Since most criteria are descriptive—and therefore abstract rather than concrete standards—there is still space for negotiation about whether a site fits the criteria or not. Such spaces were a main focus in the nomination work of local agencies, which included producing nomination documents and paperwork to be presented to the nomination committee to prove the nominated site was of "outstanding universal value."

In the application document, the government and experts proposed that Yongding Hakka tulou qualifies for numbers 1, 3, 4, 5, and 6 in the criteria list. And UNESCO experts approved the following three of these five criteria:

> Criterion 3: The tulou bear an exceptional testimony to a long-standing cultural tradition of defensive buildings for communal living that reflect sophisticated building traditions and ideas of harmony and collaboration, well documented over time.
>
> Criterion 4: The tulou are exceptional in terms of size, building traditions and function, and reflect society's response to various stages in economic and social history within the wider region.
>
> Criterion 5: The tulou as a whole and the nominated Fujian Tulou in particular, in terms of their form are a unique reflection of communal living and defensive needs, and in terms of their harmonious relationship with their environment, an outstanding example of human settlement. (UNESCO, n.d.-b)

Although the national and provincial governments were the guiding agencies in direct contact with UNESCO, the local governments on the county, town, and village levels conducted most of the on-the-ground work. In May 1998, the Yongding County government formed the Office of Tulou World Heritage Nomination Committee. The first director of the office was Hu Daixing, a local scholar who is also the director of Yongding Hakka Tulou Culture Research Association and the director of Yongding Museum. The preparation work for the nomination started with the application of Hukeng Town as a provincial historical town and the listing of some representative tulou as official government-designated national relics. A year later, the county government made a bigger step forward and sent application documents for the tulou World Heritage nomination to the higher-level authorities for review. On April 30, 2000, the Fujian Province government held a governor's office meeting and proposed the term Fujian Tulou. The proposed Fujian Tulou for World Heritage nomination included three tulou clusters and two individual tulou in Yongding and the same number of tulou clusters and individual tulou in the neighboring region in Zhangzhou.

From 2000 to 2003, the nomination of Fujian Tulou had to be postponed as the Keynes Resolution took effect. To prevent imbalance in the number of UNESCO World Heritage Sites among the members of the World Heritage Convention, the resolution stipulated that countries like China that already had sites listed could nominate only one World Heritage Site each year while those with no site listed yet could submit two to three sites for nomination. In 2004, an amendment to the resolution was made in the Twenty-Eighth World Heritage Convention held in Suzhou, China, which loosened the limitation and permitted states such as China to submit two applications each year. China was one of the countries that actively proposed this amendment. In the same year, Fujian Tulou finally got on the national Tentative List of World Heritage Nomination (*shijie yichan yubei mingdan,* 世界遗产预备名单). In 2006, the State Cultural Relics Bureau made the decision to submit Fujian Tulou to UNESCO for the 2008 World Heritage nomination. The proposal was officially approved by the State Council (Guowuyuan, 国务院) in 2007. In the summer of 2007, accompanied by experts from the State Cultural Relics Bureau, the director of Fujian Cultural Relics Bureau, an expert on architectural conservation was sent by the International Council on Monuments and Sites along with other related officials in the province to inspect and evaluate tulou for the nomination.

Considering factors such as the representativeness of tulou, the condition of tulou, and the level of local social development, three tulou clusters (Hongkeng tulou cluster, Chuxi tulou cluster, and Gaotou tulou cluster) and two individual tulou (Yanxiang Building and Zhenfu Building) in Yongding, mentioned previously, were included in the final selection of tulou by provincial governmental authorities and cultural experts for the World Heritage nomination. This group included eleven round tulou and twelve square tulou. Nine of the twenty-three tulou are in the Hongkeng tulou cluster. According to official statistics provided by the Yongding Office of Tulou World Heritage Nomination, the attempted conservation area covers 572.5 hectares (about 1,415 acres) of land in six villages. More than ten thousand people, representing over three thousand households, were directly affected by the nomination.

Since the late 1990s, the county government took the World Heritage nomination not only as a cultural opportunity for the conservation of tulou but also as a political opportunity for the popularization of Yongding and an economic opportunity for tourism development. The investment (in terms of personnel, finance, and political resources) in the nomination work was enormous. Upon the State Council's approval for the application of Fujian Tulou for nomination in January 2007, the Yongding County government announced it would "give all it had" to "secure" the success of the nomination. Nineteen high-ranking county government officials were directly involved in the work of securing the World Heritage nomination. With the leadership of the Office of World Heritage Nomination Committee, more than seven hundred staff and officers from forty-one governmental departments and organizations participated in the preparation work. The official statistics show that the accumulated spending on the tulou World Heritage nomination was RMB 2.8 billion (approximately USD 400 million) from 1998 to 2008. The major financial support was from county-level, provincial, and national governments.[2]

The application work included renovation of tulou, exhibition and archive construction, demolishment and relocation, road construction, and landscape improvement. At the suggestion of scholars and conservation experts, forty-four tulou in Yongding were selected for conservation. Wooden tablets introducing the tulou were put up in the hallways of these selected buildings to mark the protected property. The life of residents in these buildings started to be regulated by the local authorities. They were requested to keep their houses tidy and clean and were no longer allowed to raise fowl in the designated buildings like they used to do. For each building, an elder resident with a good reputation was selected as the building leader to head, coordinate, and inspect residents' work related to tulou World Heritage nomination and tulou maintenance. During rainy seasons, the building leader was also responsible for checking for leaking places in the building. The government had firefighting facilities installed in the protected tulou. Within the protected facilities, electricity would be cut off automatically in case of lightning. Stone landmarks were set up on the boundary of the conservation zones. The county government also invested in building new facilities, including tourist service centers, public restrooms, and parking lots.

In addition, to meet UNESCO requirements for application documents, the Yongding County government conducted large-scale documentation of the tulou. The content of the UNESCO World Heritage application form is an important reference for local documentation work. The application form includes geographic information and maps of the nominated site; conservation regulations and legislation; and documentation of the nominated site/item in the form of written descriptions, pictures, diagrams, and videos. Various actions were taken to meet the requirement. For instance, interviews were conducted with residents to build an oral history archive. Professionals from architectural institutes in Shanghai were hired to measure tulou, draw diagrams, and make maps with technical instruments and equipment that local residents had never seen before.

Another significant part of the nomination preparation work was to present and promote Hakka culture through media, exhibits, and other means. The county government, working with local experts, built museums in tulou for exhibitions of local culture (fig. 4.1). Various forms of media, such as radio, newspapers, television shows, and banners, were used by governmental

Figure 4.1 Zhenfu Building located in Xipian Village, Hukeng Township, Yongding County in Fujian Province. The building is a Fujian Tulou World Heritage Site and has been transformed into a tulou museum. (Photo by author)

publicity teams to introduce and promote tulou and Hakka culture. Brochures promoting knowledge of tulou, Hakka culture, and World Heritage were delivered to each household in the nominated tulou clusters to "educate" local residents. These political and cultural practices related to the tulou World Heritage nomination deeply and widely influenced and transformed local understanding of tulou and Hakka. Tulou, which was generally regarded as an out-of-date residence and a symbol of backwardness, become a source of pride for local people.

In the local government's point of view and practice, one of the most important tasks in the nomination process was the so-called environmental remediation (*huanjing zhengzhi*, 环境整治), which required the cleaning up of the environment or improving landscape at the villages selected for UNESCO World Heritage nomination. The major work of the "environmental remediation" in the process of preparing tulou for World Heritage nomination was to demolish buildings regarded as "out of place." This remediation process has had a direct impact on local daily life and social structure. In fact, demolition generated the most complaints and conflicts in the local community during the nomination process.

Zhen, a local high-ranking official who was the second and last director of the Office of Tulou World Heritage Nomination Committee, told me that demolishing new construction was the most important work performed through the local nomination process. In February 2001, the Yongding Tulou Environmental Improvement Headquarters was founded specifically to take charge of the "improvement" work. Zhen was the vice commander in chief of the headquarters. Before participating in the work of the tulou World Heritage nomination, Zhen had been the director of the Office of Demolition and Relocation and had worked on the relocation of more than twenty thousand people during the construction of the Mianhuatan Hydroelectric Power Plant, a grant project described in chapter 2. I met him in his big modern-style office with its black leather sofa set. He was an outgoing man in his fifties who was very proud of his contribution to the success of the tulou World Heritage nomination.

Zhen said that if he could use one word to summarize the preparation work of the World Heritage nomination, it would be *demolition*. The most important work for the local government had been to tear down new construction regarded as "not matching the tulou landscape." The rationale behind this was the concept of "maintaining the original/historical status of the landscape" (*xiujiu rujiu, baochi lishi fengmao*, 修旧如旧, 保持历史风貌) to rehabilitate a site or an object to return it to its condition prior to the new construction. The concept was publicly proposed by the director of the State Cultural Relics Bureau and promoted as the major working principle for those in the area of museums, cultural relics, historical sites, and heritage sites in China. Zhen claimed that historical sites had been "ruined" by social development, saying, "The tulou World Heritage Sites are selected in economically underdeveloped places in our county. Tulou in the more developed areas have long been ruined. Thus, there's no way to have World Heritage Sites in these areas." He added, "However, even in the underdeveloped places, there are some recently constructed buildings that affect the original landscape of the place. They are a blot on the landscape. Therefore, it was important to demolish them before the nomination."

In order to go back to what the village used to look like before "the development," the houses and small stores of 126 households in Hongkeng were demolished before the UNESCO nomination was submitted. In total, approximately 200,000 square feet of construction were demolished because they were "damaging the original landscape," which is to say, the landscape before the introduction of new architectural styles, techniques, and materials such as

Figure 4.2 Tangxialong, a new village next to Hongkeng for relocation.
(Photo by author)

concrete and cement in the local community. According to UNESCO conventions, local people could not build new constructions in the core zone of about 74 acres and in the buffer zone of about 178 acres in Hongkeng. Therefore, the relocation site was chosen in the adjacent village, which is dominated by another lineage with the last name of Li (fig. 4.2). Among the 126 households who need to be relocated, 101 moved to the new site outside their own village. Many of those relocated families commute between the two villages. They come back to Hongkeng for work or business during the day and go to their new home at night.

MOTIVATIONS FOR NOMINATION

During the long process of tulou nomination with all of the practices described, Yongding County experienced the change of four county leaders as well as changes and adjustments of personnel in the nomination committee six times. Nevertheless, the pursuit of the tulou World Heritage designation was a persistent effort. Why was the UNESCO listing so important to the local government? What were the motivations for the long-lasting efforts? My interview with Zhen, as one of the most important local leading figures in the nomination process, may provide some clues to answer these questions.

Zhen told me the World Heritage nomination of tulou was important based on the belief of "earning credit for the present generation and bringing benefits for future generations" (*gong zai dangdai, li zai qianqiu,* 功在当代, 利在千秋). This is one of the slogans that was used during the Yongding World Heritage nomination campaign. This is also a value commonly accepted and a slogan widely used to promote the preservation of cultural heritage by cultural practitioners in China. But it had been localized when the Yongding government used it in promoting and advocating the tulou World Heritage nomination. To convince local residents that World Heritage nomination would earn merits for the present generation, Zhen used a vernacular term, *gongde* (功德), to explain it to the people of the region. Gongde is a folk religious term meaning that the good deeds you do in your life will accumulate and benefit your afterlife and your descendants. He related tulou nomination to the ancestral worship that is deeply rooted in local people's lives and value systems:

> Tulou is a wonderful heritage created by our ancestors. We should preserve it. The best way to preserve tulou is through the World Heritage nomination. When tulou become World Cultural Heritage, they will be protected by laws such as the Convention concerning the Protection of World Cultural and Natural Heritage on the international level. On the national scale, there are national regulations for the preservation of cultural relics as well. And the Fujian provincial government released Regulations of World Cultural Heritage Preservation. We could protect our tulou through such conventions and regulations. If we do not have tulou designated as World Heritage, they won't be effectively protected. We know that Hakka people believe in loyalty and filial piety, and respecting our ancestors is the most important part of loyalty and filial piety. Thus, it is the biggest shame if we as the descendants are not able to preserve the cultural heritage that our ancestors created. This is an easier way for the people to understand [the importance of World Heritage nomination].

As implied by Zhen's words, the knowledge system of UNESCO World Heritage is foreign to the local people. Such knowledge is interpreted and mediated through local officials, scholars, and other practitioners involved in heritage work, like Zhen and Su. The meaning and significance of World Heritage is advocated by elites to the local people through the understanding and perception of local cultural and political authorities. The local authorities take the UNESCO World Heritage designation as the most effective way to achieve tulou preservation. They believe the enforcement of conventions and regulations on the legitimated heritage will provide legal protection for tulou. Furthermore, moral judgment was applied in local promotion of the tulou World Heritage nomination. Officials claimed that it would be disrespectful to their ancestors if local people did not support the nomination work of tulou, because the success of the nomination would put ancestral heritage under the protection of international and national laws and regulations. For tulou to be successfully designated as World Heritage Sites would be meritorious for the current generation, according to the government officials. They also claimed that it would benefit their descendants, which would make the current generation great ancestors to future generations. This application of local beliefs and values reflects the mediating role of local authorities. Whether local people bought the story or not, it shows local authorities' attempts to find common ground for the local residents to more easily understand the importance of tulou nomination and safeguarding and to reduce local resistance toward practices that would greatly affect their personal lives, such as the demolition of their modern houses in heritage sites.

The merit of the heritage nomination is closely associated with the economic benefit local officials think would be brought by the designation of tulou as a World Heritage Site. Zhen bluntly said that the economic opportunity brought by the World Heritage label was the most important aspect that motivated local efforts for the UNESCO World Heritage nomination. He emphasized the experts from UNESCO informed them that World Heritage is not only for conservation but also for utilization. Zhen, as one of the local officials in charge of the tulou World Heritage nomination, had the privilege of direct contacts with UNESCO experts. From these contacts, he learned a new term: *living cultural heritage*. Based on his own interpretation, he synthesized his understanding of the utilization of World Heritage and the specialist term *living cultural heritage* with his observation of the ongoing social process of modernization and urbanization in China:

> It would be such a spectacle if tulou, as well as the mode of local production and life, remain the same in one hundred years. There would be few places in the world like it at that time. The experts from UNESCO told me that tulou is living cultural heritage. It is alive rather than dead. Why is it living? There are people living in it and their life is constantly changing. Therefore, it is not only the architecture that needs to be

preserved, but it is also important to preserve the lifestyle, the mode of production, and the historical site in its original state.

When city people come to visit us, it would be like people from the countryside going to big cities like Shanghai. They would be fascinated. The urbanization process is fast. In our country, more than 50 percent of people live in cities. City people would like to visit tulou. They have no idea what tulou and rural life is like. Thus, we need to preserve farming culture like in an agriculture-dominated society. I suggest that farmers do not need to do farming any more. What should they do? We should bring corporate operation to tulou heritage sites. A corporation buys all the fields. We keep the most rural way of production. We use cattle to plow the land. That would be a great idea. Farmers would receive a salary from plowing the land for the corporation. And the corporation would buy all the farm products. It has nothing to do with farmers. We could also earn money from taking pictures. We could charge tourists ten or twenty Yuan for taking pictures with the plowman.

Even though Zhen cites "the experts from UNESCO" as the authoritative source for living cultural heritage, his understanding and interpretation of the term seems to diverge from the actual meaning of the concept. With his subtle observation of social circumstances and realization of the commercial value of heritage and traditional culture as museum-like objects representing otherness to modern tourists, Zhen saw an economic opportunity for creating a new mode of theme park that takes advantage of localness and ethnicity for the exhibition of "more authentic" community culture and daily life. Zhen fantasized about creating a fossilized agricultural community operated by a modern corporation.

Zhen's conception of a fossilized community is connected to people's search for otherness and authenticity in rural heritage-oriented touristic activities. This is frequently described as a characteristic longing that is fostered by the experience of modernity (Culler 1981). People in the modern world often believe that authenticity exists only in the past or in other regions. Zhen's vision of the touristic development echoes Dean MacCannell's (1999, 91) claim that modernity detaches local networks, histories, and folk practices from their "traditional" roots and transforms them into "cultural productions and experiences." Modern individuals develop an interest in the "real life" of others, especially those living in the countryside, as a result of their sense of having lost their own connection to a "traditional community." As rural areas and minority ethnic groups are treated as signs of the authentic past and the cultural other, agricultural tourism and ethnic tourism have become a thriving industry in China during the last two decades, a growth concomitant with the rapid urbanization and modernization happening throughout the nation during this period (Chio 2014; Donaldson 2011).

Although the fossilization of local life for tourist development is still found mainly in Zhen's imagination, tulou heritage sites like Hongkeng are indeed under the administration of local government and the management of a tourism corporation. The Committee of Yongding Tulou Conservation and Tourism Development Management (Yongdingxian tulou baohu yu lüyou kaifa guanli weiyuanhui, 永定县土楼保护与旅游开发管理委员会) is a governmental agency founded after the success of the tulou World Heritage nomination. The committee is in charge of administrative functions, such as the conservation of tourism resources, evaluation and permits for tourism-related businesses, security in heritage sites, and supervision and regulation of the local tourism market. Fujian Hakka Tulou Tourism Development Corporation (Fujiansheng kejia tulou lüyou fazhan youxian gongsi, 福建省客家旅游发展有限公司) is a state-owned corporation that manages tourism activities and development at tulou heritage sites. Its use of heritage for tourism and general economic development, to a great extent, determines not only the activities surrounding heritage tourism, such as cultural performances, production, exhibition, preservation, and conservation, but also the development path, the specific opportunities, and even the everyday life of targeted or affected local communities such as Hongkeng Village.

During the holiday seasons in 2011, more than 8,000 domestic and foreign tourists on average flowed each day into Hongkeng, a village of approximately 2,800 permanent residents. Local daily life is now subject to a constant "tourist gaze" (Urry 1990). As various aspects of local and regional folklife, along with the tulou, are staged for exhibition, presentation, and representation, the village has turned into what some tourists call "a living museum" (fig. 4.3). Local people's lives are regulated to various degrees in response to the requirements of tourism and its management. However, like the process of the World Heritage nomination, local people who are most directly and fundamentally impacted by heritage tourism activities are usually not being effectively engaged in the top-down policy formation and specific decision-making processes. Complaints emerged among local people about the government's ignorance of their requests and goals even while it targets their homes and daily life for commodification. But this does not mean that the local people and community

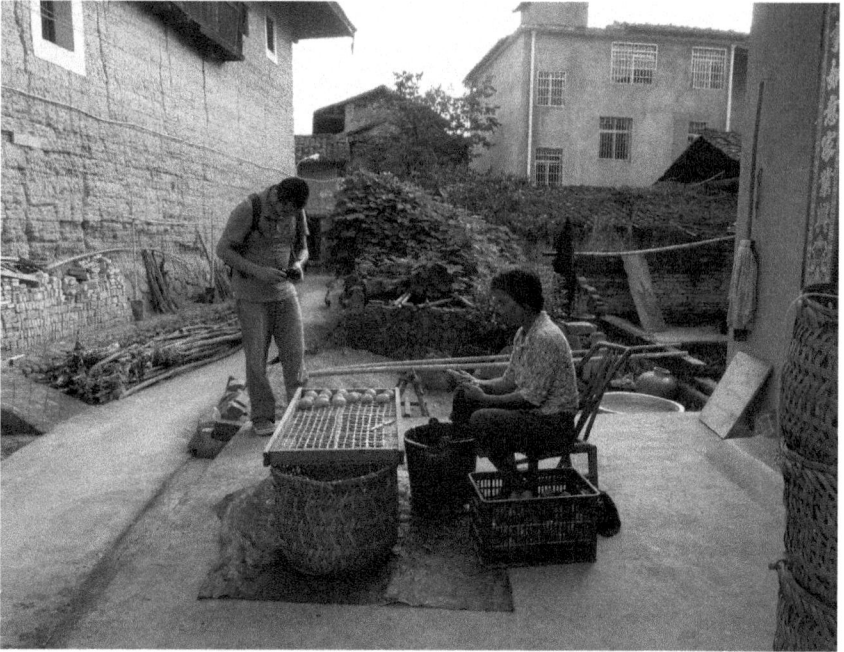

Figure 4.3 Tourist filming a community member peeling persimmons outside Qingyun Building. (Photo by author)

are all passive in this process. In the next chapters, I examine the agency of the local people and the bottom-up perspective toward the heritage process in Hongkeng Village.

CONCLUSION

Since the UNESCO World Heritage designation in 2008, tulou have been transformed from communal dwelling and property shared by a large family or families in the same lineage to heritage shared by the human beings in the world and a target for tourism development. Tulou were constructed with locally available natural materials and traditional architectural techniques to fulfill local needs of family safety and lineage development. These characteristics contributed to the case for external heritage recognition. Although tulou have been regarded as symbols of prosperity and family unity, the now highly valued houses were gradually depopulated and even abandoned in the decades before the UNESCO World Heritage List nomination. Tulou were regarded as out-of-date architecture by local people. Instead, single-family houses made of bricks and concrete, rather than those made of earth, were preferred by most locals. After tulou were designated as World Heritage Sites by the international cultural policymaker UNESCO, local people developed new conceptions of the massive vernacular architectural forms. When cultural experts and government officials "discovered" the special value of tulou and designated them as heritage, tulou were given what Barbara Kirshenblatt-Gimblett (1998a) calls "a second life." Despite the local government's authorized discourse (Smith 2006) of tulou being unique and harmonious communal dwellings created by local people's diligent and intelligent ancestors, now people move back to their tulou not because of a desire for living comfort or convenience but because of the new meanings and values brought into play by the World Heritage designation.

Local people's contact with heritage is mediated by multinational bodies, governments, organizations, scholars, and corporations with political, cultural, and economic authority and privileges. Local governments were active participants in the process of heritage nomination. The agencies with authority and power, such as the local government and intellectuals, made their own appeals and interests in relationship to their own understanding, interpretation, and exploitation of heritage as political, cultural, and economic resources. The local government took the World Heritage inscription as a political and economic opportunity and made heavy investments in the nomination work. That investment was expected to produce returns in the form of future profits. Thus, commodification became a concomitance of tulou World Heritage designation.

The local government, experts, and heritage-related national and international organizations are working together in the production of heritage, and a new discourse hegemony is formed in the process of heritage listing and designation. The complicated process and heavy investment in heritage designation makes it difficult to have bottom-up applications initiated by local residents. Rather, heritage designation heavily depends on the government's support both administratively and financially.[3] This significantly contributes to the legitimization of power structures and resource control by the involved actors. At the same time, the international and national organizations set the criteria and evaluate the heritage nomination. Local applications try to meet the official criteria and requirements. Thus, the application materials are constructed, mostly by experts in collaboration with other cultural practitioners, to meet the expectations of the application review committee and related organizations.

The discussion of World Heritage as a transnational phenomenon is conditioned by the social and cultural settings in the specific local community and, in my case, by China as a rapidly modernizing and globalizing nation-state eagerly engaged in heritage discourses and practices both nationally and internationally. At the same time, the heritage nomination process in China maintains an undertone of what Christoph Brumann and David Berliner (2016, 16) characterize as "developmentalism" in "the entrepreneurial state."

When Hongkeng, as a remote and isolated agricultural community, is resituated in the global and national system of the World Heritage program, local culture, local history, and local lives (as well as their meanings) are reinterpreted and reorganized by local authorities and exploited as a means of developing the local and regional economy. The UNESCO designation not only introduces global cultural policy to the local community but also boosts the tourism industry in the region. The cultural richness of the local community, and particularity of tulou, has been directly targeted in the recent exploitation of localness centering on tulou and Hakka culture in heritage tourism. The modern exploitation of tulou and traditional culture as socioeconomic resources is situated in the local government's ambition, ideology, and related practices that are described in chapter 2. With the opening up of this once isolated place to the world, the World Heritage List nomination and the touristic development following the UNESCO designation have become a significant driving force for the local pursuit of modernization.[4] The inscription has been used by governments and state-owned tour companies as a resource for national and international tourism marketing campaigns. Those efforts now drive vast numbers of tourists to the village. When local governmental and cultural experts started to document, repair, and display tulou and finally had them inscribed on the UNESCO World Heritage List, they paved the way for future exploitation of tulou and other forms of local resources.

NOTES

1. In 1956, Dr. Liu Dunzhen, a pioneer in Chinese architecture research, was among the first scholars to conduct academic research on Fujian Tulou. His book *Introduction to Ancient Chinese Resident Dwellings* (*Zhongguo zhu zhai gai kuan*), published in 1964, includes a description of Fujian Tulou. In 1957, scholars of architecture Zhang Buqian, Zhu Mingquan, and Hu Zhanlie published an article on tulou, "Dwellings of the Hakka in Yongding County of Fujian Province" (*Fujian Yongding ke jia zhu zhai*) in the *Journal of Nanjing Institute of Technology*. I obtained a scanned copy of the article from local Yongding scholar Su Zhiqiang. This is one of the earliest scholarly studies on tulou.

2. Statistics and information were obtained from a copy of the Conclusion for the Work of World Cultural Heritage Application for Fujian Tulou-Yongding Hakka Tulou provided by the Yongding County government on June 29, 2012.

3. During my visits elsewhere in China, mostly in Southwest China, I was informed by other people that heritage nominations and tourism development largely rely on the initiation and investment of the government.

4. *Modernization* is a term that is widely used in different contexts in China. The meaning of the concept can be ambiguous. Here modernization refers to the social, cultural, and economic transformation that leads to rapid economic growth, technology development, updated infrastructure, and high quality of life. For a comprehensive introduction of and discussion on the history and conception of modernization in China, see Soo (1989).

FIVE

—ᴍᴍ—

EVERYDAY ENGAGEMENT WITH HERITAGE PROCESS AND TOURISM ACTIVITIES

UNESCO'S PLACEMENT OF TULOU on its World Heritage List, and the touristic development that followed, has produced great social transformations among the Hakka people of Southeast China who reside in these impressive multistory communal dwellings. After Fujian Tulou received UNESCO World Heritage designation, the tulou World Heritage Sites became tourist magnets. In the context of current heritage tourism development in Hongkeng Village, many tulou are transformed into displays, performance stages, and commercial spaces. The transformation has significant impact on residents' daily routines, family communications, habitual social interactions, sense of home, and cultural presentation and representation. The enclosed buildings, whose design is touted for protecting against outside interruption and intrusion, now welcome in a large number of tourists year-round. Tourism, as an interaction-intensive industry, now brings natives into frequent direct or indirect contact with outsiders in various forms. This chapter examines the ways in which heritage-related tourism activities affect individuals and communities in everyday life and how community members participate in heritage and related tourism activities. We will see that the local people are not merely passive players in tourism. Rather, they are often active participants and are more mindful of how they represent their heritage through self-representation than some may assume. I also explore the local reshaping of space and cultural representation to illustrate the ways that heritage tourism transforms local culture and identity. Local government and community members, intentionally or unintentionally, present and represent their identity and tradition while also producing localness through the decoration and reshaping of local living space.

PRIVATE HOME AS PUBLIC COMMERCIAL SPACE

As described in chapter 4, the number of residents in tulou has decreased primarily due to people—especially the younger generation—moving out of the traditional dwellings in favor of modern single-family houses or leaving the local community for job opportunities in other places. Even though this is also true for tulou residents in the neighboring communities I visited that are not included on the World Heritage List, I specifically observed changes happening in the heritage-driven tourist sites. After tulou received World Heritage designation, the designated Fujian Tulou World Heritage Sites were targeted for tourism development, and tourism is still central to the local developmental strategy and plan. In 2011, the local tourism company set the goal of receiving two million tourists because tourism activities in heritage sites stimulate the economies of communities through employment and private businesses. With the new opportunities and possibilities, the heritage process brings some young people back to the rural area.[1]

Due to the World Heritage nomination, many of the buildings became a potential and then actual source of revenue in the tourism market. And so, young residents moved back and started tourism businesses in their buildings. Among them were my hosts in Yucheng Building, Binghan and his brother Dehan, who used to work in the cities. Binghan and his wife, Lizhen, took up the opportunity of touristic development and came back home to open a restaurant and homestay hotel. His brother became a helper in the family business. In the past, only their parents still lived in the building. Now, the building is livelier with the three generations of the family living together.[2]

One evening, after a long day of serving tourists, Binghan got the chance to take a rest and chatted with two tourists from the city. Naturally, they started talking about the differences between city apartments and tulou. Binghan said, "Living in an apartment in big cities such as Xiamen, you don't own the sky and you don't own the land," explaining that there are people living both above and under the apartment in tall buildings. Pointing to the sky and the earth, he proudly claimed that "living in tulou, the sky belongs to me and the land also belongs to me." Tulou is the resource for his daily economic income. At the same time, living in tulou also provides him with a different sense of ownership and attachment to space. Although he had bought an apartment in Xiamen, tulou provides him a more concrete sense of property ownership as the living space's direct attachment to the land and the natural environment.

The routine and pattern of life in tulou, as commercial space, are governed by touristic activities. For those who run family businesses like Binghan's family, it is overwhelmingly busy at the peak time of tourist visitation. When there are no tourists coming, people feel bored and wait for tourists to come. In one kind of social change, people have reduced their visiting of friends and clan branch relatives, as their family businesses keep them at home and waiting for tourists. Tourists usually arrive at the village between eleven o'clock in the morning and two o'clock in the afternoon. Most of them come by bus in the form of tour groups led by tour guides. During most weeks, the number of tourists increases on Friday and reaches the peak on the weekend. Usually on Monday there is a dramatic drop in the number of tourists in the village. During the year, the busiest time is summer, from July to September, when many parents take their children on vacations. Holidays and festivals also bring a large number of tourists to Hongkeng and to surrounding tulou scenic spots.

Daily life in the tulou revolves around the visitation of tourists. Every day at around five o'clock in the morning, when the sound of bells and drums from Tianhougong Temple comes to Yucheng Building, Binghan's parents and Ziwu's parents are the first people to get up in the building. They start their daily routine with the sweeping of the floor, making breakfast, boiling water, and feeding the poultry. Binghan and Lizhen get up at around eight o'clock. They usually skip breakfast and start preparing for a busy day. From around eleven o'clock, the building starts to become crowded with tourists. The noise of tourist groups makes the tulou feel and sound like being at the crowded local market. The hosts of the rice wine workshop give tours and introduce the process of making local wine in the traditional way. Binghan and his family are busy rushing to take menu orders and serving up Hakka food for the tourists. During peak times, there are more than twenty tables laid out in the restaurant, with some also on the second floor. The section on the second floor owned by Binghan's family has been remodeled as dining rooms. The house is bustling with tourists walking around, peeking into closed doors, taking pictures, and drinking tea (fig. 5.1)

Local perception and experience of "home" has greatly transformed as tulou life and tulou have become touristic spaces. On one hand, tulou became tidier and more livable due to associated renovation work for heritage nomination and tourism development. Also a factor here is the imposition of new rules—on animal husbandry inside the buildings, for instance. On the other hand, tulou become less livable as the living space is constantly disturbed by strangers, and some are eager to view the private lives of those dwelling within. This forces residents to live under "the tourist gaze." Beyond this gaze specifically, their private time—both the daily round and the seasonal one—and their living spaces are fragmented by touristic activities.

Figure 5.1 Tourists resting, drinking tea, and taking photos in Yucheng Building. (Photo by author)

The disturbance of daily family activities by outsiders is inevitable and unpredictable once a house is opened and managed as a commercial space in the tourism industry. Binghan's mother takes care of his two children when Binghan and Lizhen are focusing on their business and their customers. While the children have lunch at around noon, the adults in Binghan's family usually don't have time for lunch until two or three o'clock, when the tourists start to leave. The building becomes quieter in the late afternoon. Binghan's parents take time to clean the house, sweeping litter and collecting water and beer bottles left by tourists. Binghan often takes this time to do accounting work. Dinner is the only meal that the family has the chance to sit down and enjoy together. Even during dinnertime, there are individual travelers coming, so Binghan and Lizhen have to leave table to serve them.

Local villagers, especially of the older generation who were used to the lifestyle of a rural and remote agricultural society, have to make concerted efforts to adapt to the touristic circumstances in which they now live. One evening after dinner, Binghan's mother, who is a very easygoing and hardworking woman, complained about her current life. She told me: "In the past, when we were farming, we planted the rice and put fertilizer in the rice paddy. Then we were free to do other things. Now, all year round every day is the same. It is quite tiresome." I thought that she must have been bearing this feeling for some time, as she struggled to cope with a new life that was not the same as the life she had lived for more than fifty years. For farmers, daily activities vary as the season changes, but there's a pattern that they are familiar with; they have more control of their time and daily activities. While entering the tourism service industry, the daily touristic practice is year-round. And there's less flexibility in terms of time arrangement and daily schedule. Tourists' demands are, for commercial reasons, taken to be the priority that must be met.

Furthermore, in touristic contexts everything in daily life is on stage for display to outside spectators. The search for authenticity prompts tourists to find opportunities to peek into local people's personal lives and private spaces. "the backstage" (MacCannell 1999) is often of more interest to them. They would come to the family dining room to check whether Binghan's family eats something different for their own meal than what they serve on the menu. Unlike Binghan's family, who opened all four floors of their side of the building to guests (as they remodeled most rooms upstairs as dining room and guest room), Ziwu's extended family, who ran the rice wine workshop, only opened the ground floor of their side to visitors. The upper floors of their half of the building were their private living space. Although they had to constantly tell tourists who tried to go upstairs that upper floors were private space, they were still able to preserve more private space for family life outside of constant view. The boundary between private and public space largely depended on the type of tourism business the residents engaged in and the residents' comfort zone of sharing space with strangers.

CROSS-CULTURAL AND INTERGROUP ENCOUNTERS AND BOUNDARIES

In the process of Hongkeng villagers' increasingly frequent cross-cultural and intergroup encounters with the visitors and other agents in heritage and tourism activities, the boundary of public space and private living space has become blurred while the boundary between *us* and *other* seems to be more visible. Erving Goffman (1959) distinguished between "backstage" and "frontstage" in regard to an individual's or group's positioning in performance and life. The backstage and frontstage are often seen in tourist destinations such as the Dai communities in Southwest China studied by Sara Davis (2005). The backstage is the private area where people live their lives. Their action and communication does not need to meet the audience's expectations, and they do not have to represent a certain identity. Frontstage is where an individual or a group engages in performance as they are knowingly watched by an audience. The settings direct the action and communication. Sometimes the frontstage is arranged and made to look like it is the backstage. Between the backstage and frontstage, there are various degrees of middle ground that is either closer to the frontstage or closer to the backstage. In Hongkeng, the frontstage and backstage have been increasing blurred and intertwined in the touristic context. Space and activities in people's daily lives, even though they are not staged for an audience, become potential targets for the tourists' gaze. In some cases, public space is expanded and intrudes on the villagers' intimate lifeworlds.

While the frontstage and backstage are increasingly blurred, the villagers distinguish and shift between in-group community of the acquaintance's society and the out-group community, including tourists, the tourism agencies, and officials. In their interactions with people who are not their community members, they have frequent opportunities to self-consciously present and represent themselves. These intergroup and intercultural encounters also enhance a sense of boundary between us and other.

All of the villagers know each other and are connected in the lineage system. They share a tacit norm in their daily interaction and relationship. On a sunny afternoon in the summer, I heard some people chitchatting about an incident of fire burning the trees on the hill near the village. One of my hosts in Yucheng Building guessed the fire was probably caused by some villager who burned incense and papers during the ritual of cleaning the grave and offering sacrifices to ancestors. The villagers did not know who started the fire—not because they were incapable of finding out who it was, but because they did not want to find them. According to my host, the village cadres who were also members of the Hongkeng community did not even bother to investigate the case, because the community evaluated the incident and found it had not caused essential harm to anybody. According to Chinese law regarding environmental protection, the person, if identified and confirmed of being responsible for the fire incident, may have had to go through a legal process. In this case, the villagers chose to activate the unspoken protection mechanism within their society so that one of their community members would not have to face outside force and the legal system that would potentially change his or her life in a negative way.

Along with the new opportunities brought by the heritage-related development, new regulations and restrictions are also issued, usually by officials or those who are not a member of the community. The villagers have to navigate through those regulations. In the summer of 2012, when I visited the village, selling farm products in the village was discouraged during certain times of the day after some tourists complained on national media that the village had become a market with villagers selling farm products and other items by the roadside. In some tourists' opinion, this negatively impacted the image of the village and their traveling experience. As my host Qiyin quoted from the media broadcast, the tourists thought that the village became "too commercialized." In response to this kind of complaint, two nonlocal staff were hired by the tour company to patrol the village. One day, when I was documenting my hosts at Yucheng Building making snacks with taro and sticky rice powder, I heard a loud noise of people arguing outside the building. I went with the residents of Yucheng Building to check on what was happening. We saw the two patrolling staff members arguing with a lady in her fifties. The lady was carrying a small basket with homegrown peaches in it for sale. She stepped into Yucheng Building for shelter while arguing with the patrol staff. All the residents sided with the lady. Facing a group of villagers defending the lady, the two tour company staff decided to leave with only a verbal warning. Later that afternoon, when I walked around the village, I ran into the same lady selling peaches by the roadside where a large group of twenty tourists were surrounding her, showing great interest in her peaches and the scene of her selling peaches in the village.

Despite the more diverse and complex interpersonal relationship, the intercultural and intergroup encounter in heritage and tourism process provides villagers the extra space and opportunity for storytelling. They are more actively engaged in the introduction of local history, Hakka ethnicity, vernacular architecture, foodways, and other aspects of traditional culture. Some of the older generations, such as my informants Denghui and Fashan, can be regarded as "tradition bearers" who possess a wealth of local knowledge and are enthusiastic about sharing traditions with both the villagers and tourists not simply for the purpose of running a tourism business but mainly for increasing personal and community visibility. At the same time, the tourism industry has produced a large number of professional narrators, namely, the tour guides who introduce local culture to tourists.[3] Unlike Denghui and Fashan, who accumulate their local knowledge through their life experience and long-term informal communication and interaction with their community members, many of the tour guides are trained by the tour company and only started to introduce local culture after the development of heritage tourism in the village. Most of the tour guides are younger female members of the community. When I talked to

Liping, a tour guide and a resident of Guangyu Building, about her work, she said:

> It is challenging for me to introduce Hakka people [and our culture]. It is hard. For instance, about the use of a wok with heated charcoal during the wedding ritual, I just told the tourists that it represents people's wish for prosperity. And I don't know why people also use a bamboo tray for the ritual. Upon entering the village (with the tourists), I proudly introduce Hakka and the well-known Hakka figures, starting with the story of Hakka ancestors moving here from Central China and the guest [Hakka people] becoming the host [the dominant majority] in this region. Some tourists asked why those in Xiamen and Zhangzhou who also moved from Central China are not categorized as Hakka. And Sun Yat-Sen is from Guangdong Province; why is he a Hakka? I don't know how to answer those kinds of questions. I didn't know anything about Hakka before. I started to gradually learn about it after I joined the tourism industry. I read books, am trained by the tour company, and sometimes learn something from the tourists.

It is interesting to look at what is regarded as Hakka culture or as the part of Hakka culture that is thought to be representative of the community in local perception. Even though Liping herself, like other tour guides, is a Hakka and had been living traditional Hakka life in the village, she claimed that she knew nothing about Hakka and Hakka culture. Those lived experiences and practices are not regarded as knowledge and culture worthy of being introduced to tourists. Rather, the knowledge or narrative that is chosen, synthesized, and authorized by the experts—such as the wedding rituals (some are no longer practiced in local daily life) and the immigration history or origin of Hakka people—are formalized in the touristic narrative and become training material for local tour guides to "learn" Hakka ethnicity and culture. The history and narrative about Hakka origin used to be unknown to the villagers. It was not part of the local daily narrative. Now it has been repeated not only by local tour guides in guided tours but also by other members of the community on other occasions. The parts of Hakka culture selected by agents such as local experts and literati for the tourism narrative are regarded by the villagers as the Hakka culture that should be told. In touristic context, it has become the dominant narrative in guided group tours. Even though it's not uncommon for the local tour guides to improvise and introduce what they have learned in daily life, the cultural elements selected to be presented to outsiders and to represent Hakka people are regarded as the Hakka culture and have a higher level of performativity in the tourism narrative. Locally circulated stories and knowledge that the villagers are familiar with are usually told by senior male members of the community to individuals who show interest.

PRESENTATION AND REPRESENTATION

In the context of heritage tourism, some artifacts and cultural elements gain new meaning. They become representation of what people, both inside and outside the community, think characterize the locale. In the summer of 2011, when I was walking around to familiarize myself with the village during the first week of my stay in Hongkeng, I saw three men building a thatched shed on the way to a paved road that leads to a sightseeing pavilion on the top of a hill at the north end of the village (fig. 5.2). The thatched shed, as a scenic spot, was conceived by a design company from Shanghai that had been hired by the tour company. It was in a convenient and properly contextualized location. At the bottom of the hill were rice fields that survived from the large-scale land appropriation in the village. On another side of the hill was the Kuiju Building, a representative tulou facing the fields. Thus, the thatched shed was designed to represent the agricultural features of the village that it once was. In addition, it serves as a rest area for tourists where they can enjoy the panoramic view of the representative tulou, the Quiju Building. In 2010, the tour company finished the construction of a road paved with stone slabs going through the rice fields to the top of the hill and a sightseeing pavilion that overlooks the entire village. The thatched shed was the last part of the project.

The wooden frame of the shed, with four big solid wood columns, had been finished. The final work was to thatch the roof. A middle-aged man on the roof greeted me with a smile. He was a local carpenter in charge of the construction of the shed. The other two men were his assistants. They had been working on the thatched shed for more than half a month. The carpenter told me that he used to travel a lot to many other places year-round for work. He came back home as many ongoing landscaping projects required carpenter skills. He informed me that there used to be thatched sheds built in the village in the 1960s and 1970s. Those were very shabby sheds, hastily constructed for storing fertilizer and for farmers to rest in or to take shelter in from rain. He noted that even though the wood used was from small trees with the bark still on, rather than the big painted wood beams used now, they served their function well.

Now the farming population in the village has greatly decreased. Many villagers lost their farming lands to tourism-related construction. And many

Figure 5.2 Thatched shed under construction in Hongkeng Village.
(Photo by author)

of those who still have land did not think it was profitable to farm due to the
high cost of labor as well as seeds, fertilizer, and pesticides. Moreover, heritage
tourism turned many farmers into businessmen or service workers. And so,
the delicate thatched shed gains a new function and meaning in the current
context of heritage tourism. It is a representation of the locality as a farming
society and a place where tourists can take pictures. The design is purposefully
made by the tourism agency and a design company from a cosmopolitan city

to stress and mark the site as a traditional agricultural society even while local society is gradually turning away from agriculture and becoming more and more service industry oriented.

The conscious presentation and representation of local culture and identity began during the heritage nomination. The formal exhibit first started at Qingcheng Building, a square tulou built in 1937 next to Zhencheng Building. On each floor of the building, which has three stories in the front and four stories in the back, there are six rooms each on the front and back sides and six rooms each on the left and right sides. On the symmetrical line are the hallway, a big yard, and the ancestral hall. Currently there is only one large family living in it, which consists of three households (an old couple and their two sons). They take up only the right half of the building as their living space, of which two rooms on the first floor are converted to commercial space for selling tea and souvenirs. The other half of the building is rented to the government and has been converted into a museum themed around tulou and the Hakka people.

Crafted by local experts such as Hu Daxin, who is the director of Yongding Museum, the exhibit was organized into five sections: Introduction, the Treasure of Hakka in Fujian and Guangdong, the Wonder of Ancient Castles, Folklore and Local Traditions, and Globalized Hakka Tulou. The first and second floors show the emergence and developments of tulou, with descriptions, artifacts, models, and pictures. The third floor displays Hakka culture and traditions such as cured tobacco, Hakka folk songs, lantern parade customs, puppetry, and the local marriage ceremony.

While the whole village can be regarded as a large "open-air museum" that exhibits various types of tulou to outsiders, Qingcheng Building is a smaller museum that is contextualized by the tulou and the reality of life going on in the village. In addition, the museum in the building provides tourists with information that can help them better understand the "open-air museum," information that they would not be able to grasp in a one- or two-hour tour. Thus, the small museum in the building and the whole village as a museum formed a relationship of intercontextualization or intertextuality (if we perceive an exhibited artifact as a text) in heritage tourism activities.

While the exhibition in Qingcheng Building is crafted by local cultural experts, many other exhibitions in the village are presented by the individual agency of local residents. When the formally organized exhibition tends to display a single tulou to represent all tulou and show the Hakka ethnicity in general, the individually organized small exhibitions focus more on personal expression, individual tastes, and family glory. Due to its architectural features and local and media promotion, Zhencheng Building became one of the most popular and most commodified tulou. Most rooms, as well as the space of the passage and the hallway on the first floor, have been converted into stores. The building is open to tourists year-round. The most well-known figure in the building is Riden, a tour guide and a businessman. People who know him usually call him Aden. For the convenience of making contact with outsiders, many Hongkeng villagers have business cards. Most of the business cards, besides basic contact information, were printed with a tulou and the name of the tulou. Although Aden does not speak any foreign languages, he is one of several people in the village who have bilingual (Chinese and English) business cards. In addition, his business card also shows that he carries various titles, such as a member of Chinese People's Political Consultative Committee in Yongding County. As a famous tour guide, he charges much more than other local tour guides, and he gives tours only to important figures, such as high-ranking officials. Additionally, he offers tours only in Zhencheng Building rather than showing tourists around the whole village like most other local tour guides do.

Aden's family is one of the most prestigious families in the village. His grandfather Renshan built Fuyu Building and was one of the three brothers who made a fortune from selling tobacco and tobacco cutters. His father, Honghui, was one of the three brothers who fulfilled their father's and uncles' last wish to build Zhencheng Building. Most of Aden's brothers and sisters went to institutes of higher education and left Hongkeng. There are still more than ten households that live in the building. The expensive construction material and treatments of the building—such as colored glaze, marble, and exquisite carvings and paintings on the beams and rafters—demonstrate the wealth and the higher social status of the Lin family ancestors who built Zhencheng Building.

Including rooms inherited from his siblings, Aden's family owned one-eighth of Zhencheng Building, which consists of four sets of vertical rooms on the east side of the building. Aden tore down the partition walls between the small kitchens to make a larger space and turned the rooms into a big exhibit room. The room was labeled "Exhibit Room of Aden's Family." Entering the room, a large framed picture of Aden with the former president Hu Jingtao in Zhencheng Building was hung in the middle of the wall. Many other pictures of Aden and famous figures at Zhencheng Building were displayed on the same back wall on both sides. While the front wall was a display of Aden's personal achievements, the wall on the right side showed people Aden's glorious family. On the top was a wooden tablet with carved calligraphy of moral precepts written by Aden's father. It directs his descendants to be kind and righteous. A black-and-white photo of Aden's late father was hung under the tablet with a brief introduction of his life. As the earliest person in the village who received higher education overseas, he graduated from Waseda University in Japan and held significant social titles such as Yongding County head commissioner. Under the ancestor's portrait, twelve pictures of important members in Aden's family were proudly presented with Aden's picture in the very front.[4] These members were considered as those who glorified the family as they had received higher education and became socially outstanding in the village. The only exception is Aden himself, who wrote on a strip of paper under his photo saying that his highest degree was received from Hongkeng Elementary School. He did not feel that a lack of education was a weakness on his résumé. Instead, he is very proud of the fact of his being the most popular tulou tour guide and a very knowledgeable person about tulou despite the limited formal education he received.

The exhibit space is merged with commercial space. Along the wall by the passage were wooden cupboards that used to be places for storing cups and dishware. Now they are used for displaying tea, cigarettes, and souvenirs, three products that were most commonly sold in the village. The center of the room is occupied by a long rectangular tea table surrounded by a dozen bamboo chairs. This is where the hosts serve tea to tourists, a local rite of hospitality that has now been adopted in the new touristic circumstances as a commercial strategy of making contact with tourists and initiating conversation. With the exhibit, conversation naturally leads to stories of Aden's family and the construction of Zhencheng Building. When I first visited Aden's place in the summer of 2012, he had gone to Xiamen to attend a business conference that included a program on tulou exhibition. His nephew from Fuyu Building was hired temporarily to take care of the business at home for him. After a tour guide introduced the ancestors who built Zhencheng Building and explained the Aden family background to tourists, Aden's nephew invited them for tea. While having tea, curious tourists seized the opportunity to ask more detailed questions about Aden's family and Hakka culture as a whole. The conversation eventually ended with the tourists purchasing tea and souvenirs.

Some displays are centered on a theme/artifact that is closely related to the construction of tulou and touristic narratives. Furthermore, exhibition of local culture and family history as a new form of cultural expression is closely connected to the new form of daily economy in heritage tourism. Risheng Tobacco Cutter Workshop in the penthouse of Zhencheng Building is one such display. There is a small two-story building called a penthouse attached to each side of the main structure of Zhencheng Building. The one on the west side used to be a family school and now is a tea store. The one on the east side, close to Aden's property, is marked as Risheng Tobacco Cutter Workshop (fig. 5.3). It used to be

Figure 5.3   Risheng Tobacco Cutter Workshop at Zhencheng Building in Hongkeng Village. (Photo by author)

a family workshop where family members engaged in simple commodity production of products such as tobacco and tobacco cutters. Now it is a dwelling for a large family made up of the households of two brothers. A gate on the west side connects to the east side door of Zhencheng Building through a passage. The hosts open a door on the north side for easier access to the house. Above the door is a wooden tablet labeled "Risheng yandao fang" (Risheng Tobacco Cutter Workshop, 日升烟刀坊). *Risheng* is a registered brand name that literarily means "sun rising." On each side of the door is a plaque that was inherited from Lin's ancestors. One plaque is marked "Yongding tiaosi yanhang" (Yongding Cut Tobacco Store, 永定条丝烟) and the other one is marked "Linji yandao pifa zhongHang" (Lin's Tobacco Cutter Wholesale Store, 林记烟刀批发总行). In Qing dynasty and the Republic period, both businesses were prosperous and generated money for the ancestors of Zhencheng Building residents to build the giant building. A board is placed in front of the door that reads "Family History of Zhencheng Building."

With the Zhencheng Building, bunches of tobacco leaves, and a tobacco cutter as the background of the picture, the introduction explains that after tobacco was introduced to China from the Philippines and became popular across the county, Lin Renshan and his two brothers seized the opportunity to produce tobacco cutters. They branded their product as "Risheng." The three brothers worked hard together to make high-quality cutters. In three years, they successfully opened eighteen factories, including some in big cities such as Guangzhou and Shanghai. With their fortune made from selling cutters, the three brothers opened two schools in the village. One of them still exists and became the elementary school in the village. Besides charity work, they built Fuyu Building in 1880. Starting in 1912, the three brothers spent five years constructing Zhencheng Building. On the top of the board are characters welcoming tourists to visit the "scenic spot of tobacco cutter factory." Its designation as a "scenic spot" gives the place an authoritative sense and decontextualizes it from the rest of the environment. It sends a welcoming signal and draws tourists into the building.

The north side of the workshop belongs to Aden's elder brother, Shanhen, and the east side belongs to his younger brother, Tinghen. Shanhen has a living room on the west side of the north entrance. The small living room allows only a long tea table and a dozen bamboo chairs. The tea table (like most used by families who engage in the tourism economy) displays tea artifacts and tea culture. A big tray installed with an induction cooktop to boil water for making tea takes up most of the space on the table. Besides utensils, such as the kettle, teapot, tea strainer, and teacups, a pair of decorative toads made of a special temperature-sensitive material sits on the tea tray to test the water temperature. The color of the toads turns from brown to red when boiling water is poured on them. In front of the tea tray, seven small bronze figures are set on the table showing the process of picking tea, drying tea, sorting tea, baking tea, kneading tea, and tea tasting. An ashtray in the shape of round tulou is prepared for guests who smoke. Like many other local souvenirs, the ashtray is marked with the symbol of UNESCO World Heritage and the characters of "World Heritage, Fujian Tulou." There are six such tea tables in the small building to accommodate the large number of tourists coming to Zhencheng Building. The hosts have tried to squeeze as much public space for touristic activities as they can out of their own living space (fig. 5.4).

A big shelf separates the very small kitchen from the living room. Tea, with different packages, is displayed on the shelf. The living room also serves as a small souvenir store that sells rice wine, cigarettes, and other "local products" with the image of tulou printed on packages that are not actually locally

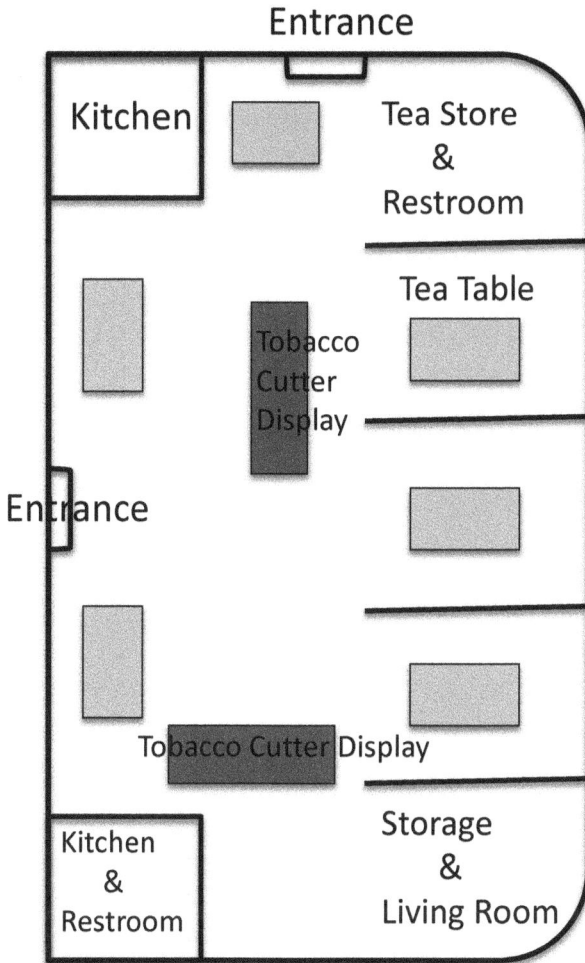

Figure 5.4 Layout of Risheng Tobacco Cutter Workshop. (Illustration by author)

produced. On the east side of the living room is a tea store. A small corner at the back has been remodeled as a restroom. There used to be three rooms on each side of the central hall on the east, but like many other families in the village, the hosts turned these rooms into open space. The front walls of the rooms and the walls between the two rooms closer to the central hall were torn down, and tourists can easily see the interior of the rooms decorated with pictures, photos, and calligraphy. Like the central halls of Zhencheng Building and Fuyu Building, a big framed painting of the house owners' ancestors Renshan and his wife was hung in the upper side of the walls in the central hall and the two extended rooms. Renshan is the twentieth generation of Lin's clan in Hongkeng and one of the three brothers who built Fuyu Building and were planning on constructing Zhencheng Building. In the picture, he wears a toga of the Qing dynasty (1644–1911). A note under the picture introduces him as a high-ranking official and a kind, generous, and patriotic man. Under the picture of Renshan is a smaller picture of Kaijie who was of the twenty-first generation of Lin's clan in Hongkeng and the great-grandfather of the current workshop owners Shanhen and Tinghen. Under the ancestors' picture is a group picture in smaller size of the current family members. The display visualizes the family lineage and its local social relationships.

The intended highlight of the building is the exhibit of tobacco cutters, which are displayed in chronological order (fig. 5.5). The exhibit, along with the remodeling of the building, started in June 2009, one year after tulou were given World Heritage status. The resident Tinghen was a migrant worker in a city before he came back home to start the tourism business in 2009. The tobacco cutters had been stored and forgotten in the building storage rooms upstairs for a long time before they were dug out and cleaned for display. The artifacts are plainly exposed to the visitors with only a piece of white paper stuck in front to mark which generation the tobacco cutter is associated with. A signboard introducing the family history contextualizes the exhibit artifacts in written form. The first generation of tobacco cutters was similar to a simple planting tool, consisting of a wooden frame and an iron blade, used by a carpenter to plane wood boards. The second generation added a knife on one side for more efficient operation. With the introduction of western technology to China in the late eighteenth century and early nineteenth century, the third generation was mostly made of iron, with gears to save energy for the operators. The fourth generation was an even more modernized small machine solely made of iron. It used a pulley and gears to control the movement of the cutter. Tourists are allowed to have hands-on experience operating the various tobacco cutters.

When tourists walk into the workshop, tour guides or the hosts introduce the tobacco cutters and link Zhencheng Building to the production of tobacco cutter technology by pointing out that the family workshop provided the financial support needed for tulou construction. The two brothers, Shanhen and Tinghen, each own half of the workshop, and work in a relationship of both cooperation and competition. They each display a set of tobacco cutters on their property. Since they sell the same products to tourists, their sales often depend

Figure 5.5 Display of tobacco cutters in Risheng Tobacco Cutter Workshop. (Photo by author)

on their relationship with local tour guides, who take tourists to the family with whom they have a closer relationship.

Although Tinghen still claims himself to be a farmer rather than a business-man, his lifestyle and daily activities are already different from a traditional farming lifestyle and activities. His family's major work in daily economy was drinking tea, chatting with tourists, making rolled cigarettes, and selling hand-made cigarettes, tea, and other products to tourists. The process of making rolled cigarettes is also a form of cultural demonstration and display. On a hot day in the summer of 2012, the temperature cooled down after an afternoon storm. Tinghen's wife was picking tea stems from a pile of tea in a woven, flat bamboo basket. Led by a local tour guide, a group of tourists walked into the workshop. After a brief introduction of the workshop by the tour guide, the visitors started to walk around and chat with the hosts. Tinghen invited them for tea. The chatting and relaxing built a more personal connection between the hosts and the tourists, which smoothed the way for further business talk.

For this group, most of the tourists ended up purchasing dried tea ready to take home at the end of the tour.

Hongkeng used to be a place for tea production. However, most tea trees were dug out for planting persimmon in the 1990s. The tea sold in the village was bought from other places in bulk with stems. Local people picked the tea stems out before selling it to tourists. Most of the tea sold also needed to be sealed in small tea packages. Like many other families in the village, there is a big freezer in Tinghen's house for tea storage and a vacuum packing and sealing machine for packaging the tea. Although the machine makes packaging much easier, Tinghen's family has to act very quickly as the tourists who want to buy tea have to leave with the whole tour group. Tinghen hired his sister, who is married to a local villager, and his sister-in-law to help with his business. They work like an assembly line. Tinghen's sister opens small bags one by one and hands them to Tinghen's sister-in-law, who uses a racking machine in front of the freezer to fill each bag with tea after weighing it on a scale. When the rack is filled with small bags of tea, it is handed to Tinghen's wife, who seals the bags with the sealing machine. Tinghen puts the small sealed tea bags in bigger package bags for the tourists.

The tea is sealed and packed on the spot as a strategy of decommodification. Like many other products sold in the village, local people try not to have commercial packages. They would not like to give tourists the impression that their products are like other commercial goods that people can get from any market. Not packing and branding the products has become a commercial strategy to advertise and positively frame their goods, as tourists find greater appeal in these so-called local products in modern commercialized society. On one hand, not packing and branding would link the goods sold in Hongkeng to preindustrial goods. On the other hand, the performance of visible local labor, in combination with the narrated history of local tea farming, repositions this nonlocal tea in a local frame. The residents' intentional act of packing goods at the scene is "a staged backstage" (MacCannell 1997) that contributes to create a sense of authenticity for tourists.

CONCLUSION

The touristic development that followed the metacultural transvaluation of tulou permits local people to make a living on the basis of newly recognized heritage assets; it also enables them to generate a new sense of cultural and identity pride and to recognize new forms of cultural expression. Now tulou are celebrated for their architectural value and for the value of expressing cultural difference and local uniqueness while concurrently being embraced by humankind as a kind of global common property. These value transformations have turned Hongkeng and other tulou World Heritage Sites into destinations. This is the basis on which local people have achieved their economic goals through the display of tulou and other expressions of local folk culture in tourism.

As tulou are put on the national and global stage as celebrated heritage and tourist destinations, the local government and local people's relationship with tulou and other forms of local culture have changed. New modes and patterns of community interaction and local cultural practices are generated under the transformative forces of heritage policy and its associated tourism. As revealed in the case of Hongkeng Village and other cases studied by folklorists, local people start to consciously and subconsciously mark and represent their traditions and identities. In Hongkeng Village, new space is constructed through local people's interaction with the world and interpretation of new situations that are different from those of their ancestors. Based on their new understanding of tulou and their own identity as Hakka in what the outside world has perceived as a traditional agricultural society, local people make modifications and even new inventions through reference to the past and recoding existing symbols to serve their current purposes. To some extent, the representation is created through the combination of outsiders' perceptions and the imaginations of the local people and place. But local people are not simply enacting a script written by outsiders; they are improvising based on the expectations of outsiders and their relationship with each other and with their own senses of local culture and history.[5]

The representation of local culture and the reorganization of living space within tulou is implicated in heritage tourism as a new form of cultural expression and a new form of daily economy in Hongkeng Village. The representation is a spacialization of time and social relationships. As in other regions impacted by folk culture–focused heritage tourism, there is a feedback relationship here. In the service industry–oriented home museum, display and reproduction is characterized by segmentation, merger, rearrangement, contextualization, decontextualization, and recontextualization of cultural elements and by the manipulation of time and space. Culture and localness seem, in their own right, to derive new meanings and values in new social settings. The knowledge of tulou construction and the history of Hongkeng as an agriculture-based society, with a foundation in tobacco cutter production and tea farming, is a resource for the cultural reproduction that has led to the transformation of a low-density living space into a high-density exhibition space (Kirshenblatt-Gimblett 1998a). When tulou as home is converted into a performance stage in a larger tourism destination, it becomes a tourist cultural space that hybridizes display, tourism narratives, everyday life, and local economy.[6] The individual and family's connection to tulou and their engagement with the heritage process makes the local community fundamental for local heritage tourism. This is why tulou and local activity can be a political resource and powerful source of leverage in the community's public engagements, as discussed in chapter 6.

NOTES

1. This phenomenon has also been observed in other places I visited in China, especially in the southwest, where cultural and ethnic tourism has been promoted to drive the development of local economy.

2. With the new opportunities and possibilities, the heritage process has attracted some young people back to the village. This phenomenon has also been observed in other places in China that I visited, especially in the southwest. These people oftentimes bring new knowledge, vision, and lifestyle back to their village. It contributes to what Yongyi Yue (2019) describes as the merge of urban and rural lifestyles in contemporary China.

3. Tour guide narrative practice in relation to vernacular architecture has been a focus for work that I have also undertaken in Beijing. See Zhang (2016).

4. It's common for people in China to display pictures of ancestors and more recent family members on the wall in their house. See Hsu (1948) for ethnographic descriptions of such displays. For research on the clan organization and the operation of Chinese family, see Lin (1948).

5. Museumification is taking place in Hongkeng and in similar places within rural China. Museumification is understood here as the transition or transformation of a living space into an idealized re-presentation of itself. Everything within such a museumified social space is considered not for its actual use in daily life but for its value as a potential museum artifact useful for presentation or representation. As Michael Di Giovine points out in his book *The Heritage-scape: UNESCO, World Heritage, and Tourism* (2009), such artifacts can be in varying degrees material, such as buildings, tools, industrial centers, markets, or more abstract things, such as ethnicity or human beings.

6. In Hongkeng Village, as elsewhere, there are many other forms of presentation, representation, and museumification underway, often pursued through commercial, corporate, nongovernmental, and governmental agencies. The reason that I pay special attention to local individual agency is that it is an emerging phenomenon of concern in current Chinese society, especially in the realm of tourism. This line of inquiry poses questions regarding what happens when self-conscious representations (such as heritage) become reestablished in turn as habitus. This matter is touched on in work by Kirshenblatt-Gimblett (1998b) and in a project that I have collaborated on with Jason Baird Jackson and Johannes Müske (2020).

SIX

—ᴍ—

TULOU AS NEW RESOURCE AND
POWER LEVERAGE

AS DEMONSTRATED IN THE PREVIOUS chapter with regard to Hongkeng Village, the public showing of private history in tulou villages transforms heritage into a resource that helps local families make a living. Heritage—a small, private, and family thing—when turned into a resource and given the World Heritage designation, can become a large, public community thing that helps the village succeed and enables the villagers to achieve some degree of agency when they encounter conflicts with outside forces. This chapter examines this transference of the private to the public, chronicling the local housing problem and illustrating the ways that World Heritage assignation and heritage tourism reconfigure broader social forces.

As previously stated, the heritage process provides opportunity for the establishment of new political power and new economic resources, and tulou World Heritage designation further changes the power relationships in local cultural, economic, and political domains. Recognition of local heritage by the outside world brings with it the attention of large corporations and governmental authorities to regulate, restrict, and generally institutionalize the daily lives of local inhabitants. However, this recognition also provides local people new resources and power to further their own interests in their interactions with more powerful outside forces. In joining the modern world through the marketing of their culture and heritage, local communities manage their symbolic and material assets while competing for resources with government agencies and tourism corporations and, in some cases, turn their cultural resource into a political one that can aid their cause in conflicts of interest.

MULTIFACETED CHANGES

While the local community is engaging in (and with) heritage practices such as listing, regulation, representation, historical re-creation, cultural commodification, and tourism management, the spatial, cultural, and social spheres of local life are impacted and transformed. The effects of the changes are multifaceted. It would be difficult to classify them into the category of pros and cons. There is no clear-cut line between the pros and the cons, as the assessment of benefits and losses is situational and relates to a range of positionalities, preferences, and histories. Those who are engaged in the tourism business would like to have as many tourists as possible visit the village, while those who do not benefit from heritage tourism do not want their life disturbed by strangers. The old are annoyed by the noise tourists make, while the young generally seem to be fine with it. When I asked my landlord Binghan his opinions and feelings, he said there are pros and cons. It sounded like he took it as a matter of course. "It's all about balancing. The only thing we can do is to try our best to maximize the pros and reduce the cons," he added.

As pointed out by other scholars, the nomination of World Heritage Sites often makes local life difficult (Hauser-Schäublin 2011; Noyes 2006). In the case of Hongkeng, the restrictions cause inconvenience in local people's lives. One of the most common problems I heard villagers complain about relates to their private property and the use of land. Before the designation, local people were persuaded or forced to tear down those structures that were identified as inconsistent with the landscape for World Heritage nomination of tulou. After the nomination, the practice of zoning within the now protected area forbids local people to have new constructions. They are not even allowed to add one more story onto their current house when their families expand. I discuss this more when examining the local housing problem and the conflicts caused by these regulations.

For the purpose of keeping the village neat and clean, local people are also told by the local authority not to let their poultry out of their cages as they used to do for generations. In another example, since more and more families can afford to own a car due to the development of the tourism industry, many people, including my landlord Binghan and Ziwu, complain about the traffic control. Private cars are not allowed to get in or out of the village between nine o'clock in the morning and four o'clock in the afternoon, during which time the major roads are often occupied by tourists.

The number of tourists traveling to Hongkeng has dramatically increased since the World Heritage designation. Before tulou was nominated for World

Heritage designation, tourism contributed to only a very small portion of the local gross domestic product. In 2004, there were around 50,000 tourists who visited tulou in Yongding County, and the ticket income was RMB 2.5 million (about USD 388,000 at the time of this publication). The number of tourists slowly grew to 62,000 with RMB 3.1 million (about USD 481,000) of ticket income in 2005. In 2006, the number of tourists almost remained the same, and the income from selling tickets to tourists was RMB 3.3 million (about USD 512,000). In 2009, one year after tulou was placed on the World Heritage List, there were approximately 600,000 tourists. In 2010, the number doubled and reached 1.2 million.[1] Each ticket is sold for RMB 90, which generates revenue of approximately RMB 108 million (about USD 16,700,000). Part of the income is distributed to the villagers. The annually distributed income, which is referred to as a "resource fee" by the villagers, provides some basic support for the low-income households in the community.

Despite all the negative effects caused by the inscription of tulou in local life, however, tulou's status as a World Heritage Site brings new opportunities and alternative lifestyles to the village. Although most people in higher-ranking positions are not from Hongkeng, some villagers got the chance to be hired by the tour company as tour guides, performers, and landscaping and cleaning staff (table 6.1). (The company also tends not to hire local villagers as security staff. The main reason is that the security staff are involved in regulating the villager's activities, and it would place the person in a difficult position if he acts against another villager's interests.)

Table 6.1. Full-time Employees of Hongkeng Scenic Spot Management Corporation. (Statistics provided by Hongkeng Scenic Spot Management Corporation on July 26, 2012.)

Office	Non-Hongkeng Employees	Hongkeng Employees	Total Employees
Vice Manager	4	1	5
Comprehensive	20	1	21
Performance	7	0	7
Security	29	0	29
Ticket	12	2	14
Electric Power Cart	17	17	34
Environmental Protection	3	6	9
Security	4	0	4
Total	**96**	**27**	**123**

Many female villagers have become local tour guides.[2] Although they are not formal employees of the company, they receive training and certification from the tour company. Their main source of income is the seller's rebate. When a tour guide takes tourists to buy tea, rice wine, and souvenirs, she or he gets up to 30 percent of the sales amount from the seller. In some cases, the local tour guide shares the rebate with the external guide who brings the tourists to the village. Local social networks and the lineage system play an important role in such tourism-based transactions. During the tour, local tour guides tend to take tourists to stores and shops run by their own family or by close lineage relatives. Small business owners who have a wide lineage connection are in a favorable position while those who don't have such connections are in a disadvantaged position.

For a long time before the tulou's nomination as World Heritage, the major income of local people came from doing handcrafts and growing a few economic crops such as tea, tobacco, and persimmon. Now, farming is no longer a major source of income for most villagers due to two factors. First, much of the already scarce farmland has been used to build tourism-related facilities such as roads and the tourist center. Consequently, people have little farmland left in the village. When looking at the pictures taken in front of Zhencheng Building and the surrounding area, it is easy to notice changes to the land. All of the land surrounding Zhencheng Building used to be paddy fields. Now that farmland has been turned into a big plaza. Second, the tourism service industry takes up a significant part of the local daily economy. Approximately one-third of the local households are actively engaged in tourism businesses such as souvenir stores, restaurants, hotels, tea stores, and the wine house. Even though there have been complaints that some young people are becoming lazy and tainted with bad habits such as drinking and gambling as they get easy money due to tourism development, many young people get the opportunity to stay with their family while making a living at home rather than working in other places. The tourism business even provides job opportunities for tulou residents' relatives and people in neighboring villages. In addition to job opportunities, the villagers are happy with the improvement of public facilities and a cleaner living environment. As far as farming goes, there is now a lack of land, a lack of labor for its pursuit, and a lack of general interest in pursuing it. Instead, people are attracted by the economic opportunities that heritage tourism has brought. Local restaurants get food delivered from outside of the village every morning. The wine sold to tourists is mostly made from sticky rice bought in the prefectural city Longyan. In fact, the making of wine is even outsourced to neighboring Hakka counties such as Dabu. Some of the villagers are directly engaged in business with tourists, but others work on supporting the new daily economy in other forms. For instance, Lin Shengyang, whose house is not on the touristic route, bought wholesale tea from Anxi in eastern Fujian Province and sells it to those who run tourism businesses in the village. At the practical level, this means that the village is an importer of agricultural and food products to an even greater degree than it ever was prior to tulou tourism.

New kinds of hierarchical social structures and social relationships emerge when heritage tourism becomes an important part of local daily life. Those who live in significant tulou or own property in these tulou, especially those who live in the Zhencheng Building, Fuyu Building, and Kuiju Building, which have been designated national cultural relics, are in an advantageous position to gain economic benefits from tourism (figs. 6.1 and 6.2). These people become economically powerful in the village. Many of the tulou engaged in tourism belong to the eldest branch of the Lin's clan in Hongkeng, which generates a new sense of pride and privilege among the branch members. Ziwu's brother, Jianwu, proudly boasted to me when we were chatting about the Lin's lineage system in the hallway of Yucheng Building, "Our branch [the eldest branch] is the biggest and most prosperous. All of the representative tulou for tourism development belong to our branch. And the three national treasure buildings also belong to our branch."

When tulou and Hakka identity are being symbolized and promoted in the heritage process, local people compete for cultural and social resources. A case that is widely noted, and gossiped about in the region, is that an old resident in Chengqi Building claimed himself to be "the tradition bearer" of tulou knowledge and that he had been telling outsiders about tulou even before tulou was placed on UNESCO's World Heritage List. When he learned that I wanted to interview somebody in the building, he proudly said he was the most knowledgeable about tulou. However, his pride was defeated when another resident, a retired teacher, was chosen by the local government to be the guide for the former president Hu Jingtao when he visited tulou in 2010 and thus became the most popular tour guide in the building. Binghan's sister Afang, who works at Chengqi Building, told me that the old resident who always thought himself to be in the authoritative position of telling outsiders about the building developed a negative attitude toward the retired teacher.

While the new social circumstances of heritage tourism may have negative impacts on local people's relationships, it also generates new kinds of cooperation. The local lineage system sets the foundation for such cooperative relationships. As previously stated, people of the same lineage branch established a sense of identification in the process of lineage interactions. Such a sense of

Figure 6.1 Fuyu Building at the north end of Hongkeng Village. (Photo by author)

Figure 6.2 Kuiju Building at the north end of Hongkeng Village. (Photo by author)

closeness and identification extends to the realm of heritage tourism. As noted, local tour guides tend to bring tourists to do business with people of their own lineage branch. It is also common that brothers join forces for the opening and managing of family business. Ziwu, who used to operate a small medicine store in the village, runs the rice wine workshop with his two brothers, who used to be migrant workers. Examples like these illustrate how heritage tourism is restructuring but also building on the local system of social relationships.

Heritage tourism brings people as well as business back to the village. In a small store next to Guangyu Building, a couple who lived in Guangyu Building made and sold malt candies and ginger candies. The store owner proudly marked the production location as Hongkeng on the package of the candies. They said their family had been making candies for more than twenty years, with the skill passed to them from the husband's father. In the past, they carried the candies to sell in different places. Now, with a consistent flow of customers, they hire some helpers and are able to open a store at home, and for the first time, they have a brand name for their products.

SITE AND INSTRUMENT OF PUBLIC NEGOTIATION

On the morning of September 27, 2012, three days before the seven-day holiday of the National Day, I was writing in my field journal in my room on the fourth floor of Yucheng Building. I heard my landlord, Binghan, coming upstairs while talking on the phone and saying: "It is closed now. The government is negotiating with them. With the previous experience of a similar situation, I am sure it will be opened soon. You can go to my restaurant and have lunch first." I was assuming Binghan was talking with a tour guide who was going to take tourists to Hongkeng. Binghan has good connections with tour guides who brought around one hundred tourists on an average day to dine in his restaurant. However, he seemed to be talking about something unusual. With people gossiping in the building about this one topic, I soon found out what was going on. Zhencheng Building was closed! Binghan was talking with a tour guide on the phone and asking her to take the tourists to his restaurant before they could maybe possibly visit Zhencheng Building and the village.

As stated, Zhencheng Building is among the three so-called national treasure buildings in the village that were most popular among tourists and also one of the most commercialized. The public space of these three representative tulou, such as the central hall and the passages of the first floor, had been rented to the state-owned tour company to be public space for touristic activities. Thus, in a normal situation, these buildings have to be open for the tourists every day. However, on that day Zhencheng Building was heavily guarded with a thick earthen wall and the closed door wrapped with an iron sheet. As Zhencheng Building was closed, the whole village became unusually quiet with few tourists. Two security guards and several staff from the local tour company were sitting in the yard in front of the building, and a tour guide was on the phone negotiating how she could deal with the tourists who had already arrived at the village but could not go into the building and visit it. At the same time, tourists who had bought entrance tickets for that day were asking for refunds.

In order to understand why the building was closed, we have to step back for a moment to understand the broader context of the housing situation in Hongkeng Village. The closed door of Zhencheng Building was a protest against the government relating to the unsettled housing problem, which had been a concern for local people since the government started the tulou World Heritage nomination in the late 1990s. As told by Zheng Dinglai, the vice director of Yongding Tourist Industry Development Committee and the director of the already dismissed Hakka Tulou World Cultural Heritage Nomination Committee following the completion of tulou nomination work, one of the most important tasks in the nomination process was to tear down recently constructed buildings as they were thought not to fit the "original landscape" of the historical site. In Hongkeng, the houses of 126 households were torn down. About one hundred households were relocated to the "new village" outside Hongkeng, with the others moving into the remaining housing stock.

After the tulou complex was successfully designated as World Heritage Sites, the local government marked the core zone and buffer zone of the heritage site in order to provide protection to the nominated area. The core zone covers the living area throughout the entire village (fig. 6.3). The buffer zone, which is meant to provide "an additional layer of protection to a World Heritage property" (UNESCO, n.d.-a), serves as a protection area that extends to the hills and mountains around the valley. The local government made the policy that there should not be new construction or remodeling in the core zone so as to protect the "historical authenticity and environment integrity" of the heritage site. For the buffer zone, new construction is not encouraged, and remodeling of existing buildings needs to be done only after a successful permit application. The regulation has been enforced strictly. During my stay in Hongkeng, a crew of security men hired by the government patrolled the village every day to make sure there was no new construction. There were several sites in the village where people tried to build houses, and the projects were stopped in the early stages of construction with the concrete house frames pulled down by the government.

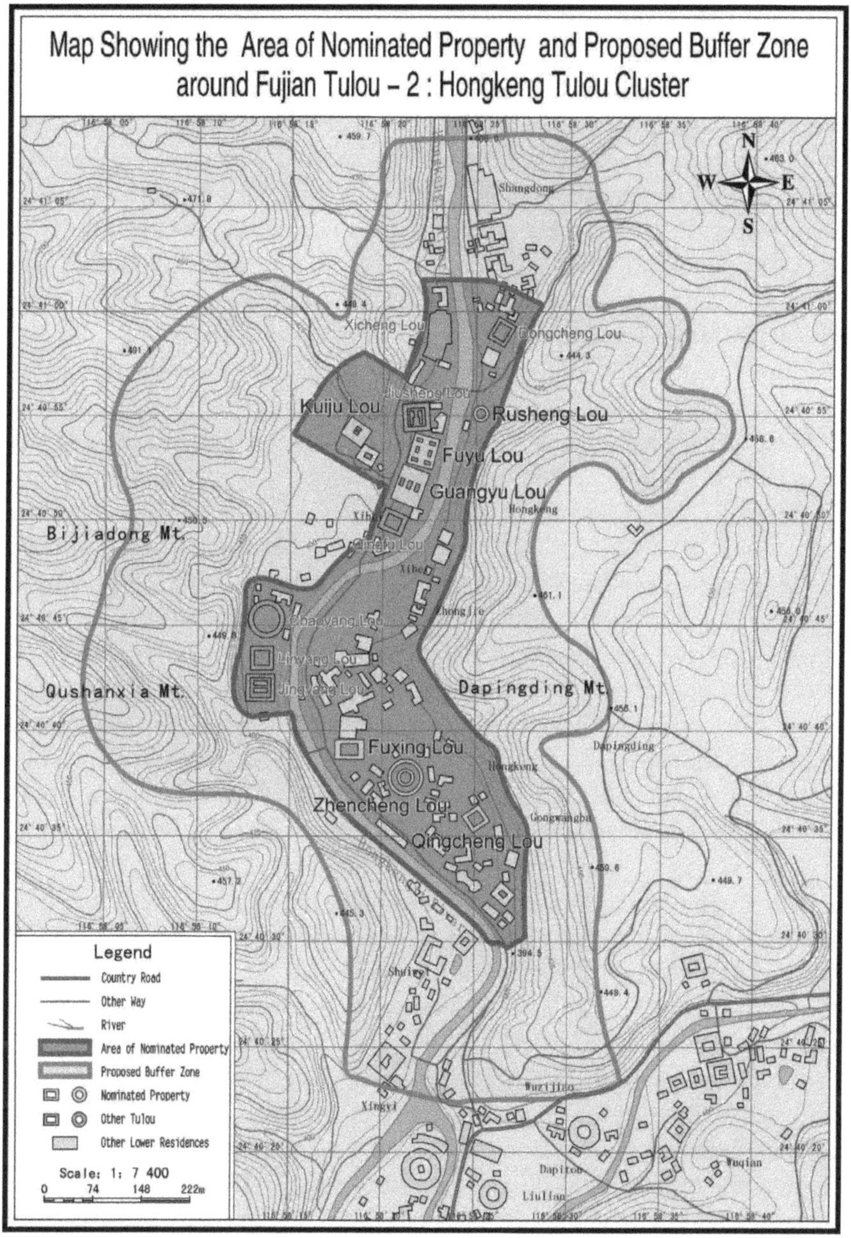

Map Showing the Area of Nominated Property and Proposed Buffer Zone around Fujian Tulou – 2 : Hongkeng Tulou Cluster

Figure 6.3   Core zone and buffer zone of Hongkeng Tulou World Heritage Site, accessed on September 27, 2023, https://whc.unesco.org/en/list/1113/maps/

In addition to the restriction on new construction due to heritage-related policy, there are other reasons for the local housing problem. As pointed out, as with Zhencheng Building, a majority of the living space in tulou has been taken up by touristic business. Most of the kitchen and dining room on the first floor have been turned into souvenir and tea stores. Many local residents in tulou such as Zhencheng Building have to cook in their small bathrooms at the inner ring of the building. Furthermore, many people who used to be migrant workers have come back to the village for the new job opportunities that heritage tourism provided—so there are simply more people in need of housing. For instance, there used to be only eight people living in Yucheng Building; in 2008, another four households came back to the building for tourism business, and so, the number of residents grew to twenty-three.

What is more, local people have a perception of tulou that differs from those of outsiders. Despite UNESCO and the government's celebration of the architectural, cultural, functional, and aesthetic values and meanings of tulou, many residents still think that tulou are no longer sufficient for modern life. The living spaces, such as the kitchens and bedrooms, are thought to be too small and too dark for comfortable life. Since many families live under the same roof, some people also complain that they do not have sufficient privacy. Although people try to maintain a harmonious relationship with each other, arguments and conflicts are almost inevitable due to how close the families live together in the same building. And many people also long for a modern bathroom, which most tulou do not have. Furthermore, some people want a quieter living environment, especially for their children, as tulou have become very noisy with the large number of tourists coming each day.

Since the building of new houses with modern facilities and appliances is forbidden in the village, local people have frequently raised the housing problem and requested that the government acquire a piece of land outside the village for them to build new houses on. However, the government kept giving the villagers the runaround rather than settling the problem. The tension between the residents and the government intensified when the local people got really annoyed by the government's attitude and lost their trust in the government to relocate them outside the heritage site. So, they started to take action.

First, some villagers signed a request asking the government to start genuine discussions and begin work on solving the housing problem. However, this request was ignored. On September 26, 2012, residents of Zhencheng Building, along with some other villagers, held a meeting and made the decision to close the gates of Zhencheng Building the next day as a form of protest. The timing of the door closing was very important; the villagers calculated the best time for placing greater pressure on the government during negotiations. It was arranged right before the National Day holiday, during which the local government predicted that more than twenty thousand tourists, on average, would visit the site and ticket sales would be more than RMB 1.5 million each day of the seven-day holiday. The residents knew that the government and the tour company could not afford to lose such a good business opportunity. In addition to the financial loss, the closed-door protest, if not resolved as quickly as possible, would also damage the long-term image and reputation of the local heritage tourism industry. The villagers calculated that the government and local tour company would compromise under such circumstances and ask to negotiate.

In this case, there was a reverse in the normal local power relationship. Since their private houses became the common heritage of all human beings, through UNESCO validation, local residents' inheritance, ownership, access to, and control over tulou has been dramatically altered. This heritage-born hegemony has intensified with the government regulating what residents can and cannot do inside and outside of their buildings. Residents are not allowed to do business in the passage. To avoid penalties, those who put their goods in the passage quickly have to collect their stuff when they are told that security men are approaching the building. Although the government and the state-owned tour company rent the shared space of the first floor in Zhencheng Building, the residents demonstrated their full ownership of the building by shutting the door to the government in protest. The residents were bound together by common interests and a sense of antagonism toward the government and the tour company, creating a united power while they challenged and negotiated with the external forces. Recalling its origins in conflict, the protesting residents safely stayed in the blockhouse-like building. Tulou once again performed its defensive function in a new social and historical circumstance. This time it was not wild animals and bandits that were barred from entry but political and corporation agencies and hordes of profitable tourists.

Around three o'clock in the afternoon, the government finally managed to arrange a face-to-face verbal negotiation with the protestors. There were three actors involved in the negotiation: the local people, the government, and the state-owned tour company Fujian Hakka Tulou Tourism Development Corporation. Each played a different role in local heritage tourism. The local government supervised and inspected the activities in the village. The tour company managed tourism activities. Local residents took part in heritage tourism by opening small businesses and presenting local culture as tour guides. Although the villagers and their daily life were fundamental for heritage tourism, in terms of political power and resource control, the government and the tour company maintained an advantageous position in the negotiation. However, the ownership of tulou and the timing of negotiation provided more leverage for the villagers than they had previously possessed.

The negotiators met the residents' representatives at the viewing stage on the second floor of Zhencheng Building. The villager's representatives stated what they wanted the government to do for them regarding the acute housing shortage in the village. An officer from the county government started his conversation by giving primacy to the villagers' interests with the statement that the purpose of developing tourism was to increase local people's income and improve their lives. Thus, he asked for the villagers' support for local tourism development. The head of the town leadership assured the villagers that the government would finish the land requisition for new house construction by the end of the lunar year (February 2013). After about one and a half hours, the negotiators finally came to some sort of agreement, and Zhencheng Building was open to tourists again.

A week later, my landlord, Binghan's father, Qiyin, told me that some governmental staff came to do a survey of how many people would like them to build a new house outside of the village. Hongkeng villagers were relieved to know there was substantial progress after the negotiation. In the immediate wake of the protests, the villagers centered on Zhencheng Building felt a sense of empowerment despite the challenges that heritage tourism brought to their lives. This form of strategic resistance was tested and proved to be effective.

The case shows us the complexity in local people's relationship with outside forces. The heritage process provides opportunity and space for the establishment of new relationships and new political power. As Laurajane Smith (2006, 7) noted, "heritage is not only a social and cultural resource or process, but also a political one through which a range of struggles are negotiated." Local people and community work on playing an active role in defining and redefining the relationship between the different agencies engaged in the heritage process. They use available resources, in this case the tulou property ownership, to (sometimes temporarily) restructure power relationships and create a world they want. Such a relationship is established not only through resistance but also through cooperation. The cooperation can be shown in the small gesture of villagers asking the tour company to sponsor a statue carved with the characters of "harmony" in front of Zhencheng Building. It can also be presented in interpersonal communication and interaction. On the afternoon of July 31, 2011, a member of the village leadership came to Yucheng Building with a proposal for a basketball event. He said that the tour company has committed RMB 10,000 to sponsor the event. The Hongkeng Villager Committee asked the township government for additional financial support. The proposal stated that conflicts between the different parties had intensified due to the development of tourism in the village. The committee organized the event as a gesture to ease the tension and improve the relationship between the different parties.

In this case, we can see that the power relationship is not always a one-way process where the powerful always holds the dominant position. It can be bi-directional, with the power not always held by one party. There is a back-and-forth balancing process that ensures the interests of all parties are considered. In combating the inequality of power and choice in this relationship, the local people's familial ownership of tulou is their major piece of leverage, and it played a vital role in reversing the power relationship through the process of negotiation.

In the past, local people have often been excluded from policymaking and have not had the opportunity to engage in local political discourse; however, the heritagization of their property has provided them with the resources, bargaining leverage, and confidence to participate in the sociopolitical process and has enabled a temporary power shift when they engage in dialogue and negotiation with other actors in the heritage regime.

CONCLUSION

The heritage tourism process has not only impacted the local landscape (exemplified by the transformation of farmland), cultural representation (as illustrated in the museumification of domestic spaces), and daily economy (as suggested by the case of locally finished and packaged tea) but also generated new problems in terms of living space and the power relationships between local people, the government, and the outside world. Tourism activities in and around heritage sites stimulate the economies of communities through employment and private business activity. As Binghan and Ziwu noted, there were "pros and cons" in this growth. The economic opportunities, new political relationships, and international policies that come with tulou heritage designation cause various tensions, conflicts, negotiations, and compromises among the different actors involved. When involved in the complex relationships among local people, the tour company, and the local government, residents find ways to use readily available resources to cope with the rapidly changing environment while negotiating with internal and external forces.

In the practice of heritage and tourism, local and indigenous people are often regarded as the powerless and passive receivers. This is partly due to local people's loss of control over cultural resources as outside agents begin to capitalize on cultural elements that belong in origin to a local community (Smith 2006). At the same time, as the case of tulou negotiation shows, the recognition of local heritage by outside forces can make local people see their own heritage as not only a cultural and economic resource but also a political one. Even as the outside forces "create" the heritage (by recognizing it as such) and work to exploit it for their own profits and political purposes, local people can sometimes use this heritage as leverage to attain what they need in their daily lives.

Tulou heritage tourism brings further complexity regarding power relationships, property ownership, and the politics of heritage. In the case of Zhencheng Building and the closed-door protest, heritage, tourism, tradition, power, and collective agency were woven into a complex network that allowed the residents a chance to strategically renegotiate everyday economics and political power in the process of socioeconomic transformation and modernization. It is not only a tactical act for the local people to temporarily reverse the power relation and achieve their justified demands but also an alternative way for them to pursue existential recognition and social and political visibility. The local people needed to demonstrate their ability to control heritage in order to gain (at least) temporary dominance in negotiating with the government over their cultural and social rights. As Smith (2006, 281) asserts, "the issue of control over heritage is political because it is a struggle over power—not only because different interests will have different and usually unequal access to resources of power, but also because heritage is itself a political resource."

To meet the current and changing needs of individuals and communities, when the time is right, local people can actively draw on available resources and adopt an effective strategy to create what Dorothy Noyes (2012, 10) calls "a practical repertoire." Under such circumstances, the residents in Zhencheng Building and their allies in the community actively and strategically used tulou as a political resource and source of power. When they shut the door of the tulou to the outside world, they declared their long-standing rights of ownership and access. In a sense, they also shut the door to the governmental legitimized power structure. The process of negotiating social and cultural change within and around the community is a continual process that the local people practice to remake and redefine both the social networks and their political position in response to the dynamic interplay between local situation and the larger socioeconomic changes.

NOTES

1. Statistics were provided by Committee of Yongding Tulou Conservation and Tourism Development Management on July 8, 2012.

2. During my visit, there were sixty-eight tour guides that registered at the tour company in Hongkeng. They have a relatively loose relationship with the company. The company provided training and shared 10 percent of the fee that a tour guide charged for each guide.

—ฑ—

CONCLUSION

IN THIS ETHNOGRAPHIC DOCUMENTATION of tulou vernacular architecture and the transformation of Hongkeng Village from an agriculture-based community to a World Heritage Site, service-oriented destination, and space of cultural display and presentation, I have studied tulou as local people's living experience. I have examined the tulou-related heritage process experienced by local residents and its transformative effect on community tradition, daily economy, and social relationship. I have also looked at how the heritage concepts and policies are understood, mediated, and practiced on the local level.

Heritage process plays a key role in the transformation of tulou and local community. Heritage has become a global phenomenon with some universal elements and features. Intergovernmental organizations such as UNESCO play an important role in the current World Heritage system. UNESCO deals with the safeguarding and listing of heritage. It is a major multinational organization through which nations around the world can legitimate heritage and claim culture. The UNESCO discourse of heritage is shadowed by forms of globalization and modernity that are considered to be destructive forces that put human heritage at risk. Even though UNESCO, headquartered in Paris, France, is far from the small village in the valley of Southeast China, its act of naming tulou heritage has had a profound impact on the village and community members' daily lives. Most notable are the mapping of the core zone and the buffer zone along with the rigid restriction of "new constructions" in the core zone as demonstrated in chapter 6.

In terms of the heritage nomination and the large-scale investment on heritage-related tourism, local residents of the tulou have little agency. The nomination process and the changes that occurred were driven by the officials on various levels. China has been an active player on both the World Heritage and the Intangible Cultural Heritage fronts in its pursuit of modernization, global recognition, and response to the changing cultural landscape. On the regional level, governments see heritage designation as a unique opportunity for local development, especially within the tourism-related service industry. Both the heritage nomination process and tourism development following the heritage designation require expertise in specific areas, heavy financial investment, and large-scale administrative support. Thus, local reliance on the government and outside agencies is commonly seen in the heritage nomination process.

Meanwhile, heritage practice is a localized process that is mediated by various agents and experienced by local people on day-to-day basis. At Hongkeng tulou World Heritage Site, heritage is interwoven into and interacts with the local cultural and social system that has remained in existence and shaped local practices for a long time. Local values, lineage systems, folk narratives, rituals, agricultural and ecological systems, and cultural policies all play a role in how heritage is perceived, received, practiced, performed, promoted, and utilized. Heritage is not the sole factor or force for community change. Examining the dynamic process of heritage interacting with and interweaving into other social forces and cultural forms provides a more comprehensive and holistic understanding of heritage and its relationship with local communities.

My documentation, contextualization, and analysis of tulou heritage and tourism in the small Hakka community in Southeast China attempts to enrich the ethnographic study of the Hakka ethnic community and the broader social, cultural, and economic circumstances characterizing contemporary rural communities in China. The ethnographic study explores the dynamic ways in which heritage impacts the traditionally agriculture-oriented and lineage-based community in rural China. It also identifies the diverse ways local people perceive and interact with national and international agencies and their heritage-related cultural policies. George Marcus and Michael Fischer (1999, 39) point out that ethnography should "capture more accurately the historical context of its subject, and to register the constitutive workings of impersonal international political and economic systems on the local level where fieldwork usually takes place ... [because] external systems have their thoroughly local definition and penetration, and are formative of the symbols and shared meanings within the most intimate life-worlds of ethnographic subjects." In this work, I find that local circumstances interact with the heritage regime in dynamic ways. Although the issue of cultural practices gets more complex when political and economic values are involved in vernacular culture in a specific community, the local social base and its social networks sets the firm foundation for individual and group creativity and cultural production. Cultural elements might be segmented, merged, reorganized, recontextualized, or even invented. However, new forms of culture evolve from what has already existed in people's daily lives. Although heritage tourism provides local people a new avenue for communications and making connections, the existing lineage-based social system in Hongkeng still plays a vital role in the new form of communication. To a certain extent, tourism businesses, as the new form of daily economy, even strengthen the lineage bonds in the village and intensify the social stratification already present in the ancestral lineage system. Local tourism narratives, performances, and exhibits are derived from local historical development and knowledge systems as these come into dialogue with national and international norms around such questions as "What belongs in an exhibition?" or "What amenities are essential to a sightseeing experience?"

My study also responds to the call to undertake focused, local ethnography through which to better understand the human impacts of heritage interventions on local communities. Such work is particularly salient in the presence of misinterpretations and misalignments of means and ends that disempower the very same communities that have stewarded the customs and cultural forms that motivate heritage policy, particularly in its conservationist guise, to begin with. Such ethnography is only adequate, in my view, if it also takes into account the ways that communities respond to the paradoxes of heritage policy and heritage tourism. As a field long concerned with problematics such as cultural authority, cultural essentialism, and the construction of tradition, folklore studies offer a particularly fruitful place from which to continue inquiring into such challenges.

When Hongkeng residents encounter heritage policies and practices, and when their remote and relatively isolated agricultural village is resituated in the global system of World Heritage designation, the local community (as well as their culture, history, and daily life) increasingly entangles with governmental management, market forces, and daily politics revolving around heritage. This entanglement engages and prompts continuous presentation, representation, and negotiation in local people's daily lives. In the realms and intersections of cultural expression, identity recognition, and daily economy, this entanglement also accelerates the changes and production of signs, meanings, and values.

The recognition of local heritage by outside forces can make local people see their own heritage as a political resource. Even as the outside forces "create" the heritage and work to exploit it for their own profits, the local people actively exercise their agency in the heritage process. The detailed exhibition of their own culture and family history in touristic context as the public showing of private history transforms heritage into a resource without handing over agency to outside forces. For them, "heritage" can be a small, private, family construct that helps the family make a living as shown in chapter 5. But "heritage" can also become a large, public community construct that helps the community succeed and keeps its members satisfied as demonstrated in chapter 6. The transference of the private to the public is made possible by the establishment of new power and resources in the heritage process and the more complex ownership of tulou. The heritage process transforms private residential homes into public economic and political resources that can be used by the residents as leverage when needed. Local residents have cultivated a strong sense of association and attachment to tulou through local narratives, rituals, ceremonies, and daily routines related to the careful construction of communal dwellings.

Tulou as a form of celebrated vernacular architecture, and more importantly, as local people's living experience sets the foundation for its value in the cultural, political, and economic spheres. Meanwhile, the issue of ownership and property becomes ambiguous and more complex when tulou is claimed to be the heritage of humankind. Tulou, the private property local residents inherited from their ancestors, at least on the symbolic level, become public property when they are listed as World Heritage Sites. The act of labeling tulou as World Heritage Sites provides governmental authorities the opportunity to reinterpret and redefine the ownership of tulou.

Heritage can refer to the ancestral inheritance over which local residents have unquestionable ownership. It can also refer to the object of cultural importance that is legitimated by an international organization as the common property of humanity. At the same time, when heritage brought tourism to the local community, the tourism entrepreneurs and enterprises assumed ownership of the heritage site for management and control. The multiple ownerships of tulou and its status of being both private and public property provide space for seemingly justified involvement and exploitation by external actors. But it also provides space for community members to negotiate with external actors and counterbalance the outside force and power. The heritage process not only leads to the dramatic reshaping of village and residential space, but it also generates new forms of social interaction and power relationships. Facing the new situation, local people have to adjust their position and adopt effective strategies to negotiate living space, social relationships, and power balances.

The concept of negotiation is of great significance in understanding intersubjectivity and social relations within the heritage process and the relationship and interaction of macro structure and the intimate world in a village and people's daily lives. The term *negotiation* contains multiple meanings and implications. The immediate understanding of negotiation might be in a diplomatic sense, such as business negotiation or peace negotiation. However, in the studies of folklore, anthropology, and sociology, negotiation is increasingly invoked in a metaphorical sense, which means carefully navigating one's way through obstacles or difficulties in either economic, ideological, political, or cultural spheres, including in the ways that these aspects of social life intersect in everyday life. The meaning of negotiation has particular theoretical implications when considered as a tool and skill that those without formal power can use to solve problems in complex circumstances and make decisions or compromises between two or more competing or incompatible interests. In the case of tulou World Heritage Site in Hongkeng Village, local residents have to negotiate complicated circumstances created by the new heritage and tourism activities to protect their own interests and balance power relationship with outside forces.

Negotiation is a social process with an interactional quality. It is a way that subjects engage and connect with other subjects. Negotiation might be considered in contrast to resistance. When a subject makes the decision to resist the people or forces on the other side of a power differential, the door for negotiation is closed. However, resistance could be a tactic taken to change the inclinations of the powerful so as to get a better deal when entering negotiations. For example, in the summer of 2012 when the local government refused Hongkeng villagers' request for solving the housing problem through building more relocation houses, the villagers locked the gates of major tulou to the tourists and to the government officials so that the state-owned tour company would lose a huge financial profit during the tourism peak in the so-called golden week of the National Day holiday. Thus, the villagers were able to force the local government to compromise in the negotiation process and agree to the local people's demands on the condition that tulou residents promise to open the gates.

On the literal level, negotiation means verbal dialogue between involved parties to reach agreements, solve problems, or pursue other kinds of common interests. This kind of social engagement is human interaction in verbal forms. An example of diplomatic negotiation is Stefan Groth's (2012) study of cultural property debates in the World Intellectual Property Organization. The study, applying sociolinguistic approaches, focuses on the "communicative patterns and strategies" that different actors employ in international negotiations regarding cultural property. The communicative manifestations such as the language used, the positions posed, the strategies applied in the texts, and talks of negotiations are carefully examined in the institutional ethnographic work to reveal the communicative dynamics and patterns of international deliberation on intellectual property of traditional culture.

Although the meaning and implication of negotiation in the verbal and linguistic sense are not excluded in the discussion of local interaction with various involved actors, my study of the relationship between heritage tourism and local community applies the term in the more metaphorical sense. As Laurajane Smith (2006, 4) states, "heritage is about negotiation—about using the past, and collective or individual memories, to negotiate new ways of being and expression identity." My study of tulou heritage and tourism analyzes the dynamic ways local people negotiate cultural and social changes and explore how heritage and tourism provides them an extra space for the negotiation of tradition, identity, power, social relations, and everyday life and economy. They work on deciding how much of the backstage to show in their house. They direct and guide the "tourist gaze" in an intentional way. The heritagization and self-representation practiced by community and individuals in Hongkeng Village and the neighboring communities present both the top-down and the bottom-up heritage processes.

As one of the stakeholders in heritage and tourism activities, local people's relationships with their living world and the outside world have dramatically changed since tulou and Hakka culture were transformed into symbolic capital for and in heritage tourism. They encounter new social settings when their living spaces, culture, and own embodied selves are put on stage on a daily basis. Willingly or unwillingly, they are brought into a new global and market system. Under such circumstances, local people reinterpret their culture, redefine their position in the world, and adjust their social relationships. Although they are often regarded by political and cultural authorities as passive cultural objects that need to be protected and regulated due to an outgrowth of external authorities' understandings of heritage, they actively participate in the process of identity expression, cultural production, and meaning making through the reshaping and display of their living space. When it is necessary, they take action to make their voices heard, and their demands are sometimes fulfilled as a result of specific tactics and strategies. While the ownership of tulou and broader understandings of heritage remain ambiguous in local Fujian Hakka society, the local people are not ambiguous in terms of claiming their ownership and rights when the time is auspicious or justified for them to make such claim.

The term *tactics* developed by Michel de Certeau might be the closest concept to my use of negotiation in terms of meanings and implications. Rather than interpreting ordinary people as noncreative passive recipients of traditions, symbols, and other cultural forms and practices, de Certeau uses *tactics* to refer to the actions that subjects take to reclaim autonomy from external power, to reappropriate their culture in everyday situations, and even to subvert what the agents of authority seek to impose on them. De Certeau (2011, 38) explains the term in more detail: "Tactics are procedures that gain validity in relation to the pertinence they lend to time—to the circumstances which the precise instant of an intervention transforms into a favorable situation, to the rapidity of the movements that change the organization of a space, to the relations among successive moments in an action, to the possible intersections of durations and heterogeneous rhythms, etc."

The notion of tactics celebrates the adaptive capacity of ordinary people and ordinary practices in everyday life. De Certeau sets *tactics* in contrast to *strategy*, which means top-down exercise of disciplining power, systematic planning, or classified knowledge on the macro level. Tactics are momentary "victories of the weak over the strong" (De Certeau 2011, 481) through micro devices, actions, and procedures.

My use of *negotiation* shares the interactional and subversive feature of tactics but understands negotiation as less defensive and opportunistic in nature. Negotiation is constantly active and even planned manipulations of power relationships. It is an active response to outside forces and constantly adjusting one's position in response to the environment, as what the tulou residents at the World Heritage Sites have been doing. In the context of the ongoing process of social transition and transformations that occur both locally and nationally, the concept and practice of negotiation reveal the broader implications of the roles that local communities and individuals play and the ways that local people participate in the social process of the Chinese state.

BIBLIOGRAPHY

Authors' names are represented here as they appear on the title page of their publications, with a comma indicating when a name has been inverted for alphabetization by surname. Names that were represented on the title page with surname first do not require a comma.

An Deming 安德明. 2016. "Feiwuzhi wenhua yichan baohu zhong de shequ: Hanyi, duoyangxing ji qi yu zhengfu liliang de guanxi 非物质文化遗产保护中的社区:涵义、多样性及其与政府力量的关系" [Community in ICH safeguarding: Meaning, diversity and relationship with the state power]. *Xibei minzu yanjiu*, no. 4: 74–81, 123.

An, Deming, and Lihui Yang. 2015. "Chinese Folklore since the Late 1970s: Achievements, Difficulties and Challenges." *Asian Ethnology* 74 (2): 273–90.

Bamo Qubumo 巴莫曲布嫫. 2008. "Feiwuzhi wenhua yichan: Cong gainian dao shijian 非物质文化遗产: 从概念到实践" [Intangible cultural heritage: From concept to practice]. *Minzu yishu*, no. 1: 6–17.

———. 2015. "Cong ciyu cenmian lijie feiwuzhi wenhua yichan—Jiyu <gongyue> 'liangge zhongwenben' de fenxi 从语词层面理解非物质文化遗产—基于《公约》'两个中文本'的分析" [Literary interpretation of ICH—Analysis of the "two Chinese versions" of the Convention]. *Minzu yishu*, no. 6: 63–71.

Bamo Qubumo and Zhang Ling/UNESCO 巴莫曲布嫫,张玲.联合国教科文组织. 2016. "Baohu feiwuzhi wenhua yichan lunli yuanze 保护非物质文化遗产伦理原则" [Ethnical Principles for Safeguarding Intangible Cultural Heritage]. *Minzu wenxue yanjiu* 34 (3): 5–6.

Brumann, Christoph, and David Berliner, eds. 2016. *World Heritage on the Ground: Ethnographic Perspectives*. New York: Berghahn.

Chan, Selina Ching. 2005. "Temple-Building and Heritage in China." *Ethnology* 44 (1): 65–79.

Chao Gejin 朝戈金. 2016. "Lianheguo jiaokewen zuzhi <baohu feiwuzhi wenhua yichan lunli yuanze>: Yidu yu pingzhi 联合国教科文组织《保护非物质文化遗产伦理原则》:绎读与评骘" [UNESCO ethnical principles for safeguarding intangible cultural heritage: Interpretation and comment]. *Neimenggu shehui kexue (hanwenban)* 37 (5): 1–13.

Chen Jinguo 陈进国. 2002. "Shisheng shisi: Fengshui yu Fujian shehui wenhua bianqian 事生事死: 风水与福建社会文化变迁" [Serving life and death: Fengshui and sociocultural transformation in Fujian]. PhD diss., Xiamen University.

Chen Zhiqin 陈志勤. 2010. "Feiwuzhi wenhua yichan de chuangzao yu minzu guojia rentong—Yi 'dayu jidian' weili 非物质文化遗产的创造与民族国家认同—以 '大禹祭典"为例'" [The invention of intangible cultural heritage and nation-state identification—The case of "Dayu Ceremony"]. *Wenhua yichan* 2:26–36.

Chio, Jenny. 2014. *A Landscape of Travel: The Work of Tourism in Rural Ethnic China*. Seattle: University of Washington Press.

Cohen, Myron L. 1968. "The Hakka or 'Guest People': Dialect as a Sociocultural Variable in Southeastern China." *Ethnohistory* 15 (3): 237–92.

Constable, Nicole, ed. 1996. *Guest People: Hakka Identity in China and Abroad*. Seattle: University of Washington Press.

Culler, Jonathan. 1981. "The Semiotics of Tourism." *American Journal of Semiotics* 1:127–40.

Davis, Sara. 2005. *Song and Silence: Ethnic Revival on China's Southwest Borders*. New York: Columbia University Press.

De Certeau, Michel. 2011. *The Practice of Everyday Life*. Oakland: University of California Press.

Di Giovine, Michael A. 2009. *The Heritage-Scape: UNESCO, World Heritage, and Tourism*. Lanham: Rowman and Littlefield.

Donaldson, John A. 2011. "Tourism: Joyous Village Life." In *Small Works: Poverty and Economic Development in Southwestern China*, 102–29. Ithaca: Cornell University Press.

Fan Lili 方李莉. 2016. "Youguan 'cong yichan dao ziyuan' guandian de tichu 有关' 从遗产到资源'观点的提出" [The proposal of the notion concerning "from heritage to resource"]. *Yishu tansuo* 30 (4): 59–67.

Freedman, Maurice. 1971. *Chinese Lineage and Society: Fukien and Kwangtung*. New York: Humanities Press.

Fleming, E. McClung. 1974. "Artifact Study: A Proposed Model." *Winterthur Portfolio* 9:153–73.

Foster, Michael Dylan. 2011. "The UNESCO Effect: Confidence, Defamiliarizaion, and a New Element in the Discourse on a Japanese Island." *Journal of Folklore Research* 48 (1): 63–107.

Foster, Michael Dylan, and Lisa Gilman. 2015. *UNESCO on the Ground: Local Perspectives on Intangible Cultural Heritage.* Bloomington: Indiana University Press.

Gao Bingzhong 高丙中. 2017. "'Baohu feiwuzhi wenhua yichan gongyue' de jingshen goucheng yu Zhongguo shijian 《保护非物质文化遗产公约》的精神构成与中国实践" [The spirit of the Convention for the Safeguarding of Intangible Cultural Heritage and its practice in China]. *Zhongnan minzu daxue xuebao (renwen shehui kexue ban)* 37 (4): 56–63.

General Office of the State Council 国务院办公厅. 2005. "Guowuyuan bangongting guanyu jiaqiang woguo feiwuzhi wenhua yichan baohu gongzuo de yijian 国务院办公厅关于加强我国非物质文化遗产保护工作的意见" [The General Office of the State Council's View on Enhancing the Work of ICH Safeguarding]. http://www.gov.cn/zhengce/content/2008-03/28/content_5937.htm.

Glassie, Henry. 2000. *Vernacular Architecture.* Bloomington: Indiana University Press.

Goffman, Erving. 1959. *The Presentation of Self in Everyday Life.* Garden City: Doubleday Anchor Books.

Groth, Stefan. 2012. "Negotiating Tradition: The Pragmatics of International Deliberations on Cultural Property." In *Göttingen Studies in Cultural Property*, vol. 4. Göttingen: Göttingen University Press.

Guo Yimei and Li Bingkuai 郭一梅,李炳奎. 2019. "'Sichouzhilu Zhongguo feiwuzhi wenhua yichan he minzu diqu fupin chenggo zhan' liangxiang Lianheguo Weiyena banshichu '丝绸之路中国非物质文化遗产和民族地区扶贫成果展'亮相联合国维也纳办事处 (英文)" ["Silk Road, Chinese ICH and Achievements of Poverty Alleviation in Ethnic Communities" exhibition opens at the UN Office in Vienna (English)]. *China's Ethnic Groups*, no. 3: 32–35.

Han Chenyan and Gao Bingzhong 韩成艳,高丙中. 2020. "Feiyi shenqu baohu de xianyu Shijian: Guanjian gainian de lilun tantao 非遗社区保护的县域实践:关键概念的理论探讨" [Practice of ICH community safeguarding on the county level: Theoretical discussion on key concepts]. *Zhongyang minzu daxue xuebao (zhexue shehui kexue ban)* 47 (3): 70–77.

Hafstein, Valdimar T. 2004. "The Politics of Origins: Collective Creation Revisited." *Journal of American Folklore* 117 (465): 300–15.

———. 2007. "Claiming Culture: Intangible Heritage Inc., Folklore©, Traditional Knowledge™." In *Prädikat "Heritage": Wertschöpfungen aus kulturellen ressourcen*, edited by Dorothee Hemme, Markus Tauschek, and Regina Bendix, 75–100. Berlin: Lit Verlag.

Hauser-Schäublin, Brigitta. 2011. *World Heritage Angkor and Beyond: Circumstances and Implications of UNESCO Listings in Cambodia.* Göttingen: Göttingen University Press.

He, Xiaoxin, and Jun Luo. 2000. "Fengshui and the Environment of Southeast China." *Worldviews* 4 (3): 213–34.

Hsieh, T'ing-yu. 1929. "Origin and Migrations of the Hakkas." *Chinese Social & Political Science Review* (Beijing), no. 13: 202–27.

Hsu, Francis L. K. 1948. *Under the Ancestors' Shadow: Chinese Culture and Personality*. London: Routledge and Kegan Paul.

Hu Daxin 胡大新. 2006. "*Yongding kejia tulou yanjiu* 永定客家土楼研究" [Study of Yongding Hakka Tulou]. Beijing: Zhongyang wenxian chubanshe.

Huang Hanmin 黄汉民. 1994. *Fujian tulou* 福建土楼 [Fujian Tulou]. Taiwan: Hansheng chuban gongsi.

———. 2009. *Fujian tulou—Zhongguo chuantong minju de guibao* 福建土楼—中国传统民居的瑰宝 [Fujian tulou—Treasure of Chinese vernacular architecture]. Beijing: Joint Publishing.

Jackson, Jason Baird, Johannes Müske, and Lijun Zhang. 2020. "Innovation, Habitus, and Heritage: Modeling the Careers of Cultural Forms through Time." *Journal of Folklore Research* 57 (1): 111–36.

Jiang Jinbo 江金波. 2004. "Weilongwu—Zui ju daibiaoxing de kejian jianzhu wenhua jingguan 围龙屋—最具代表性的客家建筑文化景观" [Weilongwu—the most representative Hakka architecture cultural landscape]. *Lingnan wenshi*, no. S1: 19–20.

Kirshenblatt-Gimblett, Barbara. 1995. "Theorizing Heritage." *Ethnomusicology* 39 (3): 367–80.

———. 1998a. *Destination Culture: Tourism, Museums, and Heritage*. Berkeley: University of California Press.

———. 1998b. "Sounds of Sensibility." *Judaism* 47 (1): 49–78.

———. 2006. "World Heritage and Cultural Economics." In *Museum Frictions: Public Cultures/Global Transformations*, edited by Ivan and Corrine Kratz Karp, 161–202. Durham: Duke University Press.

Koehn, Alfred. 1954. "Harbingers of Happiness. The Door Gods of China." *Monumenta Nipponica* 10 (1/2): 81–106.

Kong Qingfu and Song Junhua 孔庆夫,宋俊华. 2018. "Lun Zhongguo feiwuzhi wenhua yichan baohu de 'minlu zhidu' jianshe 论中国非物质文化遗产保护的'名录制度'建设" [On the construction of 'nomination system' of ICH safeguarding in China]. *Guangxi shehui kexue*, no. 7: 205–12.

Lin, Yueh-hua. 1948. *The Golden Wing: A Sociological Study of Chinese Familism*. New York: Oxford University Press.

Liu Aihua 刘爱华. 2016. "Chenzhenhua yujing xia de 'xiangchou' anfan yu minsu wenhua baohu 城镇化语境下的'乡愁'安放与民俗文化保护" [Nostalgia and folk culture protection in the context of urbanization]. *Minsu yanjiu*, no. 6: 118–25, 160.

Liu, Hsiu-Mei 刘秀美. 2005. *Liudui kejia diqu wuxingshi yu yanggongdun* 六堆客家地区五星石与杨公墩 [Five stars rocks and base of Yanggong in Liuduai Hakka area]. Taipei: SMC Publishing.

Liu Rui and Yang Yunyun 刘锐, 阳云云. 2013. "Kongxincun wenti zai renshi—Nongmin zhuwei de shijiao 空心村问题再认识—农民主位的视角" [Re-accessing of the issue of hollow village from the subjective perspective of the peasants]. *Shehui kexue yanjiu*, no. 3: 102–108.

Liu Xianjue 刘先觉. 2009. "Chengshi jianshe yu yichan baohu zhijian de maodun yuqi jiejue celue 城市建设与遗产保护之间的矛盾及其解决策略" [The conflicts between urban construction and heritage preservation and the strategic solutions]. *Zhongguo kexue* 39 (5): 825–29.

Liu Xiaochun and He Yixin 刘晓春,贺翊昕. 2021. "Huaxing, gongxiang yu yiyi zai shengchan—Qiangui bianjie fanxiang qingnian 'huigui difang' de shijian 唤醒、共享与意义再生产—黔桂边界返乡青年'回归地方'的实践" [Awakening, share, and the reproduction of meaning—The local practice of the youth returning to the borderland between Guizhou and Guangxi]. *Guangxi minzu daxue xuebao (zhexue shehui kexue ban)* 43 (2): 83–91.

Liu Yansui and Liu Yu 刘彦随,刘玉. 2010. "Zhongguo nongcun kongxinhua wenti yanjiu de jinzhan yu zhanwang 中国农村空心化问题研究的进展与展望" [The development and future of the study on the hollow issue in rural China]. *Dili Yanjiu* 29 (1): 35–42.

Luo Xianglin 罗香林. 1975. *Kejia yanjiu daolun* 客家研究导论 [An introduction to the study of Hakka]. Taipei: Guting shuwu.

Luo, Yu. 2020. "Safeguarding Intangible Heritage through Edutainment in China's Creative Urban Environments." *International Journal of Heritage Studies* 27 (1): 1–16.

Ma Qianli 马千里. 2017. "Feiwuzhi wenhua yichan qingdan bianzhi zhong de shequ canyu wenti 非物质文化遗产清单编制中的社区参与问题" [Issues of community participation in the ICH nomination system]. *Minzu yishu*, no. 3: 70–76.

Ma Zhonglin 马仲林. 2021. "Guanghe Kangjiacun yidi fupin banqian jumin de shengji bianqian yanjiu 广河康家村易地扶贫搬迁居民的生计变迁研究" [Study on the life of people relocated due to poverty alleviation in Kangjia Village, Guanghe]. MA thesis, Xibei Minzu University.

MacCannell, Dean. 1973. "Staged Authenticity: Arrangements of Social Space in Tourist Settings." *American Journal of Sociology* 79 (3): 589–603.

———. 1999. *The Tourist: A New Theory of the Leisure Class.* Berkeley: University of California Press. First published 1976 by Schocken.

March, Andrew L. 1968. "An Appreciation of Chinese Geomancy." *Journal of Asian Studies* 27 (2): 253–67.

Marcus, George E., and Michael M. J. Fischer. 1999. *Anthropology as Cultural Critique: An Experimental Moment in the Human Sciences.* Chicago: University of Chicago Press.

"Meizhoushi qidong kejia welongwu shenyi gongzuo 梅州市启动客家围龙屋申遗工作" [Meizhou City Initiated Hakka Weilongwu Heritage Normination]. 2016. *Kejia wenbo*, no. 1: 6.

Nitzky, William. 2013. "Community Empowerment at the Periphery? Participatory Approaches to Heritage Protection in Guizhou, China." In *Cultural Heritage Politics in China*, edited by Tami Blumenfield and Helaine Silverman, 205–32. New York: Springer.

Nowicka, Magdalena. 2007. "Mobile Locations: Construction of Home in a Group of Mobile Transnational Professionals." *Global Networks* 7 (1): 69–86.

Noyes, Dorothy. 2006. "The Judgment of Solomon: Global Protections for Tradition and the Problem of Community Ownership." *Cultural Analysis*, no. 5: 27–56.

———. 2012. "The Social Base of Folklore." In *A Companion to Folklore*, edited by Regina Bendix and GalitHasan-Rokem, 13–39. Maldon: Blackwell.

Nyíri, Pál. 2006. *Scenic Spots: Chinese Tourism, the State, and Cultural Authority.* Seattle: University of Washington Press.

Oakes, Tim. 1998. *Tourism and Modernity in China.* New York: Routledge.

Peng Zhaorong 彭兆荣. 2006. *Bianji zuqun: Yuanli diguo biyou de keren* 边际族群: 远离帝国庇佑的客人 [Marginalized Ethnic Group: The Guests Away from the Protection of the Empire]. Hefei: Huangshan shushe.

———. 2008. *Yichan: Fansi yu chanshi* 遗产: 反思与阐释 [Heritage: Reflection and Interpretation]. Kunming: Yunnan jiaoyu chubanshe.

Phillips, Carolyn. 2013. "The Kitchen God of Chinese Lore." *Gastronomica* 13 (4): 22–31.

Poon, Pauline. 2009. "The Cultural Meaning of Hakka Architecture in Hong Kong and Guangdong." *Journal of the Royal Asiatic Society Hong Kong Branch* 49:21–55.

Qian Chen and Yin Peiru 钱程, 尹培如. 2014. "Jiyu baohu yuanze de Fujian tulou hangtu yingjian jishu yanjiu 基于保护原则的福建土楼夯土营建技术研究" [Study on the rammed earth construction technique of Fujian tulou from preservation perspective]. *Huazhong jianzhu* 32 (4): 7–10.

Schein, Louisa. 2000. *Minority Rules: The Miao and the Feminine in China's Cultural Politics.* Durham: Duke University Press.

Shi Rei and Wei Xiaoli 石蕊, 魏晓丽. 2021. "Yimin banqiancun tudi paohuang de xingcheng jizhi—Yi Gansusheng Baiyinshi Jinyuanxian D cun weili 移民搬迁村土地抛荒的形成机制—以甘肃省白银市靖远县D村为例" [The formation mechanism of disused land in relocated village—The case of D village in Jingyuan County, Baiyin City, Gansu Province]. *Shanxi nongye daxue xuebao (shehui kexue ban)* 20 (5): 55–61.

Smith, Laurajane. 2006. *Uses of Heritage.* London: Routledge.

Soo, Francis. 1989. "China and Modernization: Past and Present a Discussion." *Studies in Soviet Thought* 38 (1): 3–54.

Tang Xiaoqing 汤晓青. 2014. "Feiwuzhi wenhua yichan baohu yu chuancheng zhong difang jingying de diwei yu zuoyong 非物质文化遗产保护与传承中地方民俗精英的地位与作用" [The status and role of local folk elites in the safeguarding and transmission of intangible cultural heritage]. *Wenhua yichan yanjiu* 4:3–14.

Tian, Yu, and Liang Hongzhang 田宇,梁宏章. 2020. "Guangxi feiwuzhi wenhua yichan minglu tixi jianshe zhong de wenti yu duice 广西非物质文化遗产名录体系建设中的问题与对策" [The issues in the construction of ICH listing system in Guangxi and some suggestions]. *Guangxi minzu daxue xuebao (ziran kexue ban)* 26 (2): 36–41.

Tian Zhaoyuan 田兆元. 2009. "Guangzhu feiwuzhi wenhua yichan baohu Beijing xia de minsu wenhua yu minsuxue xueke de mingyun 关注非物质文化遗产保护背景下的民俗文化与民俗学学科的命运" [The fate of folk culture and folkloristics in the context of ICH safeguarding]. *Henan Shehui kexue*, no. 3: 4–7.

———. 2014. "Jingji minsuxue: Tansuo rentongxing jinji de guiji—jianlun feiwuzhi shengchanxing baohu de benzhi shuxing 经济民俗学: 探索认同性经济的轨迹—兼论非遗生产性保护的本质属性" [Economic folklore studies: An exploration of the trajectory of identification economy and the essential features of ICH productive safeguarding]. *Huadong shifan daxue xuebao (zhexue shehui kexue ban)* 2: 88–97.

United Nations Educational, Scientific, and Cultural Organization (UNESCO). n.d.-a. "Convention Concerning the Protection of the World Cultural and Natural Heritage." World Heritage Convention. http://whc.unesco.org/en/conventiontext/.

———. n.d.-b. "The Criteria for Selection." World Heritage Convention. http://whc.unesco.org/en/criteria/.

———. n.d.-c. "Fujian *Tulou*." World Heritage Convention. http://whc.unesco.org/en/list/1113.

———. n.d.-d. "Text of the Convention for the Safeguarding of Intangible Heritage." Intangible Cultural Heritage. https://ich.unesco.org/en/convention.

Urry, John. 1990. *The Tourist Gaze: Leisure and Travel in Contemporary Societies*. London: Sage.

Wang dan 王丹. 2020. "Feiwuzi wenhua yichan fuwu minzu diqu jingzhun fupin de shijian moshi 非物质文化遗产服务民族地区精准扶贫的实践模式" [The mode of ICH practice contributing to poverty alleviation in ethnic communities]. *Zhongnan minzu daxue xuebao (renwen shehui kexue ban)* 40 (5): 64–69.

Wang, Yu. 2008. "Naturalizing Ethnicity, Culturalizing Landscape: The Politics of World Heritage in China." PhD diss., Department of Cultural Anthropology, Duke University.

Watson, James. 2004. "Presidential Address: Virtual Kinship, Real Estate, and Diaspora Formation: The Man Lineage Revisited." *Journal of Asian Studies* 63 (4): 893–910.

Weaver, Martin E., and Frank G. Matero. 1997. *Conserving Buildings: A Manual of Techniques and Materials*. Hoboken, NJ: Wiley.

Wen Chunxiang 温春香. 2006. "Fengshui yu cunluo zongzu shehui 风水与村落宗族社会" [Fengshui and village clan society]. MA thesis, Fujian Normal University.

Williams, Michael Ann. 1991. *Homeplace: The Social Use and Meaning of the Folk Dwelling in Southwestern North Carolina*. Athens: University of Georgia Press.

Xia Yuanming 夏远鸣. 2011. "Shi ping 'Zhongguo pinglun' zhong liang pian guanyu kejia yuanliu de wenzhang 试评《中国评论》中两篇关于客家源流的文章" [Review on Two Articles on the Origin of Hakka in *China Review*]. *Jiaying xueyuan xuebao (zhexue shehui kexue)* 29 (12): 20–23.

Xie Huazhang 谢华章. 2004. "Fujian tulou hangtu banzhu de jianzao jiyi 福建土楼夯土版筑的建造技艺" [The construction skills of Fujian rammed earth buildings]. *Zhuzhai kejia*, no. 7: 39–42.

Xie Zhongyuan 谢中元. 2014. "Guonei xinnongcun jianshe yu feiwuzi wenhua yichan baohu guanglian yanjiu shuping 国内新农村建设与非物质文化遗产保护关联研究述评" [Literature and discussion of the ICH safeguarding research relating to the construction of new countryside in China]. *Guangxi shehui kexue*, no. 3: 54–59.

Xue Li 薛力. 2001. "Chengshihua beijing xia de 'kongxinchun' xianxiang jiqi duice tantao—Yi Jiangsusheng weili 城市化背景下的 '空心村' 现象及其对策探讨—以江苏省为例" [The phenomenon of "hollow village" in the context of urbanization and the countermeasure for dealing with it—The case of Jiangsu Province]. *Chengshi guihua*, no. 6: 8–13.

Yang Chunhua and Yao Yiwei 杨春华, 姚逸苇. 2021. "Hewei 'nongcun kongxinhua'—Yige jiegouhua de gainian fenxi shijiao 何谓 '农村空心化'?—一个结构化的概念分析视角" [What is "hollow village?"—The conceptual analysis from structural perspective]. *Nongcun jingji*, no. 7: 79–86.

Yang Lihui 杨利慧. 2020. "Shequ qudong de feiyi kaifa yu xiangcun zhenxing: Yige Beijing jinjiao chengshihua xiangcun de fazhanzhilu 社区驱动的非遗开发与乡村振兴:一个北京近郊城市化乡村的发展之路" [Community-driven ICH development and the revival of the countryside: The development path of an urbanized village in the suburb of Beijing]. *Minsu yanjiu*, no. 1: 5–12, 156.

You, Ziying. 2020. *Folk Literati, Contested Tradition, and Heritage in Contemporary China: Incense Is Kept Burning*. Bloomington: Indiana University Press.

Yu Ying 余英. 1997. "Kejia jianzhu wenhua yanjiu 客家建筑文化研究" [Cultural study on Hakka architecture]. *Huanan ligong daxue xuebao (ziran kexue ban)*, no. 1: 14–24.

Yue, Yongyi. 2019. "Disciplinary Tradition, Everyday Life, and Childbirth Negotiation: The Past and Present of Chinese Urban Folklore Studies." In *Chinese Folklore Studies Today: Discourse and Practice*, edited by Lijun Zhang and Ziying You, translated by Wenyuan Shao and Yuanhao Zhao, 27–61. Bloomington: Indiana University Press.

Zhang, Juwen, and Xing Zhou. 2017. "Introduction: The Essentials of Intangible Cultural Heritage Practices in China: The Inherent Logic and Transmission Mechanism of Chinese Tradition." *Western Folklore* 76 (2): 133–49.

Zhang, Lijun. 2016. "Performing Locality and Identity: Rickshaw Driver's Narratives and Tourism." *Cambridge Journal of China Studies* 11 (1): 88–103.

———. 2019. "Institutional Practice of Heritage-Making: The Transformation of Tulou from Residences to UNESCO World Heritage Site." In *Chinese Folklore Studies Today: Discourse and Practice*, edited by Lijun Zhang and Ziying You, 146–76. Bloomington: Indiana University Press.

Zhang, Qiaoyun. 2020. "Intangible Cultural Heritage Safeguarding in Times of Crisis: A Case Study of the Chinese Ethnic Qiang's 'Cultural Reconstruction' after the 2008 Wenchuan Earthquake." *Asian Ethnology* 79 (1): 91–113.

Zhao Di 赵迪. 2021. "Pailou mantan 牌楼漫谈 [On Pailou]." *Minyi* (4): 81–86.

Zhao Yuanhao 赵元昊. 2021. "Cong Huaiyang ninigou guancha minjian yishu cunxu de wenhua shengtai 从淮阳泥泥狗观察民间艺术存续的文化生态" [Examining the cultural ecology for the existence and continuity of folk art through the case of Huaiyang clay sculpture dog]. *Minjian wenhua luntan*, no. 1: 97–106.

Zheng, Jing. 2013. "The State Army, the Guerrillas, and the Civilian Militias: Politics and the Myth of the Tulou, 1927–1949." *Traditional Dwellings and Settlements Review* 24 (2): 51–64.

Zhou Xing 周星. 2012. "Feiwuzhi wenhua yichan baohu yundong he Zhongguo minsuxue—'Gonggong minsuxue' zai Zhongguo de kenengxing yu weixianxing 非物质文化遗产保护运动和中国民俗学—'公共民俗学'在中国的可能性与危险性" [ICH safeguarding and Chinese folklore studies—The possibilities and risk of 'public folklore' in China]. *Sixiang zhanxian* 38 (6): 1–8.

———. 2013. "Minjian xinyang yu wenhua yichan 民间信仰与文化遗产" [Folk religion and cultural heritage]. *Wenhua yichan* 2:1–10.

Zhu Gang 朱刚. 2017. "'Yidai yilu' chanyi yu feiwuzhi wenhua yichan baohu de guoji hezuo '一带一路'倡议与非物质文化遗产保护的国际合作" ["Belt and Road" initiative and the international cooperation of ICH safeguarding]. *Xibei minzu yanjiu*, no. 3: 39–47.

INDEX

Aden, 127
authenticity, 110, 120

backstage, 75, 120–21, 134
bagua, 68
Belt and Road Initiative, 12
Binghan, 5, 6, 80, 89, 117–20, 138, 143, 149
boundary, 121
buffer zone, 144. *See also* UNESCO: zoning practice

Chengqi Building, 53–55, 141
clan, 80–86, 92n2. *See also* lineage
commercial strategy, 134
commodification, 17n4, 19, 94, 111, 113
core zone, 144. *See also* UNESCO: zoning practice
cultural industry, 11

daily economy, 117–34, 143
decommodification. *See* commercial strategy
defamiliarization, 98–99
demolition, 105–6, 138
developmentalism, 108–10, 114

developmental plan, Yongding County, 24–27. *See also* developmentalism
documentation, cultural, 104–5. *See also* institutionalization, cultural
Dongcheng Building, 56
Dongsheng Building, 59–60
door god, 62, 73n7

empowerment, 148
empty nesters, 27
environmental remediation, 105–6
ethnography, 153–54
exhibit, 126, 128–34
exorcism, 62. *See also* ritual

Fengcheng Town, Fujian Province, 7, 20–30, 42
feng shui, 53. *See also* folk beliefs
fertility ceremony, 78
five elementary stones (*wuxingshi*), 58–59
folk beliefs, 53, 57–58
fossilized community, 109–10
frontstage, 120–21
Fujian *Tulou*, 49–50, 99, 102
Fuyu Building, 89

gazetteer, 8, 22, 33, 40, 45
genealogy, 76, 80–81
General Office of the State Council,
     12–13
Gongde. See moral judgement
Guangyu Building, 76–80

habitus, 97–98
Hakka, 40–42, 44, 46n8, 50, 59, 71, 71n4,
     93, 95, 99, 108, 116, 140, 157; ancestors,
     53; cultural display, 126; culture, 48,
     97–98, 104–5, 114, 123–24, 128, 157;
     community, 153; custom, 55; ethnic-
     ity, 49, 122; folk belief, 37–38, 58; food,
     118; history, 47; identity, 135, 141
Hakka Exposition Park, 42–44
Hakka Tulou Research Association, 97
heritage: heritagization, 9; intangible
     cultural, 10, 11; investment on nomi-
     nation, 104; motivation for nomina-
     tion, 107–11; policy, 12, 13; regulation
     and restriction, 104, 107, 122, 138, 144;
     safeguarding, 12, 17n4
heritage fever, 11
heritage making, 94
heritage movement, 11
hollow village, 27
Hongkeng Village, 1–6, 7–9, 15, 19, 42,
     45, 48, 56, 116, 137, 144, 152, 156, 157;
     agriculture and economy, 38–40; col-
     lectivization of agriculture, 38; desti-
     nation, 28; early industrialization of,
     38; entrance, 3, 16n2; family name, 33;
     folksong, 35–36; geographical feature
     of, 30, 35; layout of, 29, 84; location,
     28; migration, 33–35; new space, 135;
     social-geographical unites, 83–85;
     social organization, 34, 91; social
     transformation, 7, 9; spatial and social
     correlation, 29–30
housing problem, 144–46

Hu Daxin, 49, 126
Hukeng Township, Fujian Province, 20,
     24, 26, 28, 102

infrastructure, 17, 23–25
innovative urbanization, 12
institutionalization, cultural, 13, 49, 93,
     98–107
intercontextualization, 126
International Council on Monuments
     and Sites, 103
intertextuality, 126

Jiang Jicheng, 55
Jiang Linghong, 53

Keynes Resolution, 102–3
kitchen god, 65–66

land transfer, 27
legitimization, 113
Li Dingdu, 58
lineage, 44, 80–86
Lin Family's Ancestral Temple, 31–32
Lin Fucheng, 81–82
Lin Guya, 3–5
Lin Kaijie, 132
Lin Renshan, 127
Lin Shengyang, 141
Liping, 122–23
Liu Dunzhen, 95
Lizhen, 4, 6, 75, 78, 80, 90, 117–20
Longyan, Fujian Province, 1, 20, 24,
     28, 140

MacCannell, Dean, 110
materiality, 72
Matsu, 37–38
Mianhuatan Hydroelectric Power Sta-
     tion, 23
migration, 89

Minnan tulou, 50
modernity, 110
moral judgement, 108–9
Mout Wuyi, 99
museum, 104–5; living, 111; open-
    air, 126
museumification, 15, 136n5, 149

nationalism, China, 11
negotiation, 147–48, 155–58
new countryside construction, 9, 12, 17n5

Office of Yongding World Heritage
    Nomination Committee, 7, 102, 106
otherness, 110–11

positionality, 138
poverty alleviation, 12
power leverage, 147–48, 150
power relationship, 137, 148–49
presentation, cultural, 7, 52, 104–5, 116,
    124–26
property ownership, 117, 147, 155

Qingcheng Building, 36, 82, 126
Qiyin, 75–76, 122

red couplets, 88, 92n4
reform and open up, 10
relocation, 106–7
representation, cultural, 124
repurpose of farmland, 140
Riden. See Aden
Risheng Tobacco Cutter Workshop,
    128–34
ritual, 53, 57–59, 60–62, 75, 77–78

Shanhen, 131
Shizhong Township, Fujian Province, 88
Shu Ziqiang, 96
social conflict, 144–47

social relationship. See lineage
souvenir, 131
State Council, 103
State Cultural Relics Bureau, 103
Strategy, 158

taboo, 56
tactics, 158
Taoism, 57–59, 68. See also folk beliefs
thatched shed, 124–25
Tianhougong Temple, 37–38
tobacco, 7, 22, 32, 37, 39, 55, 68, 89,
    126–36, 140
tooth tablet, 76, 92n1
tourism, 10, 26; commodity, 133–34;
    competition, 141–42; corporation,
    111; development management, 111;
    employees, 139–40; impact of, 118–20;
    infrastructure of, 124; narrative, 123;
    professional narrators, 122; revenue,
    139; routine in daily life, 117–19; tour
    guide, 151n2
tourist gaze, 111–12, 120
tradition bearer, 122
transvaluation, 117, 134–35
tulou: central hall, 58, 61, 63, 66–67,
    68–71, 76–77, 86–89, 131; construc-
    tion material for, 57; depopulation, 88,
    90; design, 63–67; entrance hall, 86;
    ethnic association of, 49; favorable
    location, 53; featured form, 48–49;
    foundation of, 59–60; function of,
    63–66; gate of, 62; housing allotment,
    86; land acquisition for, 53; life in the
    past, 89–90; location of, 51; occu-
    pancy rate of, 90; ordinary form, 48;
    renovation, 104; shape of, 56; as social
    ordering system, 81–83; stove in, 62;
    terminology, 48–49; tiling of, 61; wall
    construction of, 60
Tulou Visitor Center, 2

unattended children, 27
UNESCO, 49–52, 72, 104–5, 116, 144, 152; convention, 9, 10, 11, 107; as cultural policy maker, 113–14; experts, 109–10; knowledge system, 108; selection criteria and vocabulary, 90, 93–95, 100–102; validation, 147; zoning practice, 138
urbanization, 10, 12, 26–27, 44, 109–11
utilization, 109–10

vernacular architecture, 47, 155. See also *tulou*

wedding ceremony, 76–80
weiwu (enclosed house), 5–52
weilongwu (enclosed dragon house), 50–52
World Heritage Site designation, 1, 7, 93, 100, 107–9, 113, 116, 117, 137. *See also* UNESCO
Wu Yifu, 56
Wuyun Building, 54–55

Yanggong, 57–59, 61, 67, 86. *See also* folk beliefs
Yang Junsong. *See* Yanggong
Yongding County, Fujian Province, 20–28; agriculture, 22; geographical feature, 22; origin of, 20
Yongding County Department of Publicity, 98
Yongding County Land and Resource Bureau, 27
Yongding Hakka Friendship Association, 97
Yucheng Building, 4–6, 63, 89

Zhang Jiangu, 58
Zhengcheng Building, 68–71, 127–34, 143–44
Zheng Dinglai, 100, 106, 108–9
Zhengzhong Building, 88
Ziwu, 4, 36–40, 80–83, 120, 143, 149

LIJUN ZHANG is Assistant Professor of Folklore at George Mason University. She is editor with Marsha MacDowell of *Quilts of Southwest China* (IUP, 2016) and with Ziying You of *Chinese Folklore Studies Today: Discourse and Practice* (IUP, 2019).

FOR INDIANA UNIVERSITY PRESS

Allison Chaplin *Acquisitions Editor*
Anna Garnai *Editorial Assistant*
Sophia Hebert *Assistant Acquisitions Editor*
Samantha Heffner *Marketing and Publicity Manager*
Brenna Hosman *Production Coordinator*
Katie Huggins *Production Manager*
Darja Malcolm-Clarke *Project Manager/Editor*
Dan Pyle *Online Publishing Manager*
Michael Regoli *Director of Publishing Operations*
Jennifer Witzke *Senior Artist and Book Designer*

www.ingramcontent.com/pod-product-compliance
Lightning Source LLC
Chambersburg PA
CBHW040256290326
41929CB00052B/3433